AF714797

Advanced Testing of Systems-of-Systems 1

Advanced Testing of Systems-of-Systems 1

Theoretical Aspects

Bernard Homès

iSTE

WILEY

First published 2022 in Great Britain and the United States by ISTE Ltd and John Wiley & Sons, Inc.

ISTE Ltd
27-37 St George's Road
London SW19 4EU
UK

www.iste.co.uk

John Wiley & Sons, Inc.
111 River Street
Hoboken, NJ 07030
USA

www.wiley.com

Library of Congress Control Number: 2022943899

British Library Cataloguing-in-Publication Data
A CIP record for this book is available from the British Library
ISBN 978-1-78630-749-1

Contents

Dedication and Acknowledgments

Inspired by a dedication from Boris Beizer[1], I dedicate these two books to many very bad projects on software and systems-of-systems development where I had the opportunity to – for a short time – act as a consultant. These taught me multiple lessons on difficulties that these books try and identify and led me to realize the need for this book. Their failure could have been prevented; may they rest in peace.

I would also like to thank the many managers and colleagues I had the privilege of meeting during my career. Some, too few, understood that quality is really everyone's business. We will lay a modest shroud over the others.

Finally, paraphrasing Isaac Newton, If I was able to reach this level of knowledge, it is thanks to all the giants that were before me and on the shoulders of which I could position myself. Among these giants, I would like to mention (in alphabetical order) James Bach, Boris Beizer, Rex Black, Frederic Brooks, Hans Buwalda, Ross Collard, Elfriede Dustin, Avner Engel, Tom Gilb, Eliahu Goldratt, Dorothy Graham, Capers Jones, Paul Jorgensen, Cem Kaner, Brian Marick, Edward Miller, John Musa, Glenford Myers, Bret Pettichord, Johanna Rothman, Gerald Weinberg, James Whittaker and Karl Wiegers.

After 15 years in software development, I had the opportunity to focus on software testing for over 25 years. Specialized in testing process improvements, I founded and participated in the creation of multiple associations focused on software testing: AST (Association of Software Tester), ISTQB (International Software

1 Beizer, B. (1990). *Software Testing Techniques*, 2nd edition. ITP Media.

Testing Qualification Board), CFTL (Comité Français des Tests Logiciels, the French Software Testing committee) and GASQ (Global Association for Software Quality). I also dedicate these books to you, the reader, so that you can improve your testing competencies.

Preface

The breadth of the subject justifies splitting this work in two books. Part I, this book, covers the general aspects applicable to systems-of-systems testing, among them the impact of development life cycle, test strategy and methodology, the added value of quality referential, test documentation and reporting. We identified the impact of various test levels and test techniques, whether static or dynamic.

In the second book, we will focus on project management, identifying human interactions as primary elements to consider, and we will continue with practical aspects such as testing processes and their iterative and continuous improvement. We will also see additional but necessary processes, such as requirement management, defects management and configuration management. In a case study, we will be able to ask ourselves several useful questions. We will finish this second book with a rather perilous prospective exercise by listing the challenges that testing will need to face in the coming years.

These two books make a single coherent and complete work building on more than 40 years of experience by the author. The main aspect put forward is the difference between the traditional vision of software testing – focused on one system and one version – and the necessary vision when multiple systems and multiple versions of software must be interconnected to provide a service that needs to be tested thoroughly.

August 2022

1

Introduction

1.1. Definition

There are many definitions of what a system-of-systems (or SoS) is. We will use the following one: “A system-of-systems is a set of systems, software and/or hardware, developed to provide a service by collaborating together, by organizations that are not under the same management”. This simple definition entails challenges and adaptations that we will identify and study.

A system-of-systems can be considered from two points of view: on the one hand, from the global systemic level (we could take the image of a company information system) and, on the other hand, from the unitary application system (which we may call a subsystem, application system or application, software-predominant equipment or component). We will thus have at the upper level a system-of-systems that could be a “information system” that is made of multiple systems that we will call subsystems. For example, a company may have in their information system one accounting system, a CRM, a human resource management system, a stock management system, etc. These different systems are most likely developed by different editors and their interaction provides a service to the company. Other examples of systems-of-systems are air traffic systems, aircrafts and satellite systems, vehicles and crafts. In these systems-of-systems, the service is provided to the users when all subsystems work, correctly and quickly exchanging data between them.

Systems-of-systems, even if they are often complex, are intrinsically different from complex systems: a complex system, such as an operating system, may be developed by a single organization (see Figure 1.1) and thus does not respond exactly to the definition as the subsystems are developed under the same hierarchy. The issue of diverse organizations and directions (see Figure 1.2) implies technical, economic and financial objectives that may diverge between the parties and thus multiple separate systems creating, when put together, a system-of-systems. A more exhaustive description is presented in ISO21840 (2019).

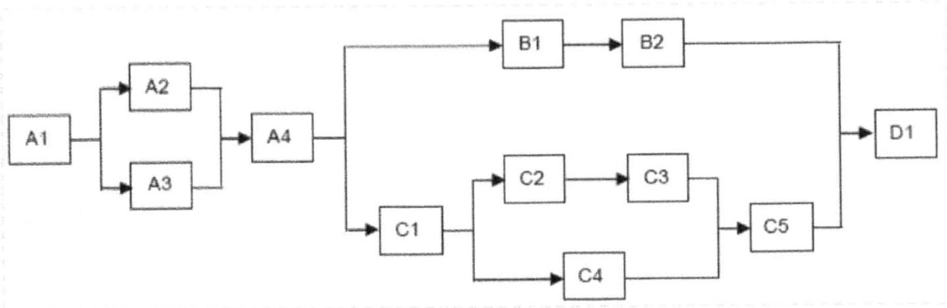

Figure 1.1. *Complex system*

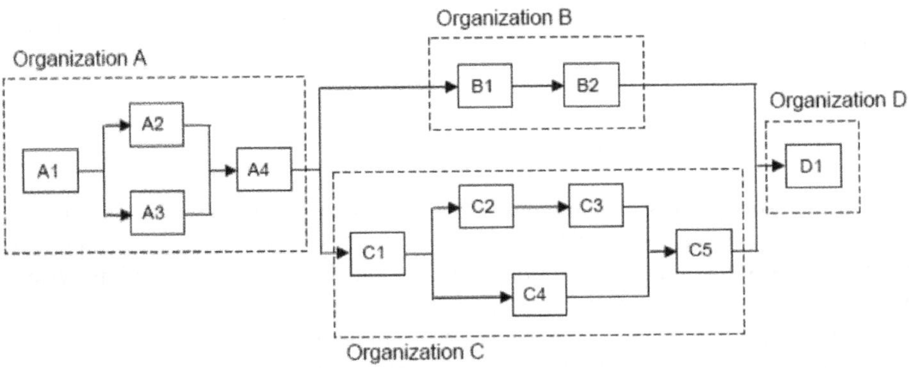

Figure 1.2. *System-of-systems*

Usually, a system-of-systems tend to have:

– multiple levels of stakeholders, sometimes with competing interests;

– multiple and possibly contradictory objectives and purposes;

– disparate management structures whose limits of responsibility are not always clearly defined;

– multiple life cycles with elements implemented asynchronously, resulting in the need to manage obsolescence of subsystems;

– multiple owners – depending on subsystems – making individual resource and priority decisions.

It is important to note that the characteristics differ between systems and systems-of-systems and are not mutually exclusive.

1.2. Why and for who are these books?

1.2.1. *Why?*

Why a book on the testing of systems-of-systems? Systems-of-systems are part of our everyday life, but they are not addressed in software testing books that focus only on one software at a time, without taking into account the physical systems that are required to execute them, nor the interactions between them that increase the difficulty and combinatorial complexity of testing. To ensure quality for a system-of-systems means to ensure for each subsystem (and sub-subsystem) the quality of the design process for each of these systems, subsystems, components, software, etc., that make them up.

Frequently, actors on a system-of-systems project focus only on their own activity, resecting contractual obligations, without considering the requirements of the overall system-of-systems or the impact their system may have on the system-of-systems. This focus also applies when developing software to be used in a company's information system: the development teams seldom exchange with the teams in charge of support or production. This slowly changes with the introduction of DevOps in some environments, but the gap between IT and business domains remains large.

As more projects become increasingly complex, connected to one another in integrated systems-of-systems, books on advanced level software testing in the frame of these kinds of systems become necessary.

Most books on software testing focus on testing one software for one structure, where those that define requirements, design the software and test it are in the same organization, or – at least – under the same hierarchy. These are thus a common point for decisions. In a system-of-systems, there are at least two sets of organizations: the client and the contractors. A contractual relationship exists and directs the exchanges between these organizations.

Many specific challenges are associated with these contractual relationships:

– Are requirements and specifications correctly defined and understood by all parties?

– Are functionalities and technical characteristics coherent with the rest of the system-of-systems with which the system will be merged?

– Have evolutions, replacements and possible obsolescence been considered for the whole duration of the system-of-systems being developed?

In a system-of-systems, interactions with other systems are more numerous than in a simple system. Thus, the verification of these numerous exchanges between components and systems will be a heavier load than for other software. In case of defect, it will be necessary to identify which party will have to implement the fixes, and each actor will prefer to reject the responsibility to others. These decisions may be influenced by economic factors (it may be cheaper to fix one system instead of another), regulatory factors (conformance may be easier to demonstrate on one system instead of another), contractual or technical (one system may be simpler to change than another).

Responsibilities are different between the client and the organization that executes the development. The impact is primarily felt by the client, and it is up to the development organization to ensure the quality of the developments.

The increase in the complexity of IT solutions forces us to envisage a more efficient management of specific challenges linked to systems-of-systems to which we are increasingly dependent.

1.2.2. *Who is this book for?*

Design of software, systems and systems-of-systems requires interaction between many individuals, each with different objectives and different points of view. The notion of "quality" of a deliverable will vary and depend on the relative position of each party. This book tries to cover each point of view and shows the major differences between what is described in many other books – design and test of a single software application – with regard to the complexity and reality of systems-of-systems. The persons who could benefit from reading this book are as follows:

– design organization project managers who must ensure that the needs of users, their customers and their clients are met and therefore that the applications, systems and systems-of-systems are correctly developed and tested (i.e. verified and validated);

– by extension, the design organization we will have assistant Project Managers, who will have to ensure that the overall objectives of the designing organization are correctly checked and validated, especially taking into account the needs of the users – forever changing given the length of systems-of-systems projects – and that the evidence provided to justify a level of quality is real;

– customer project managers, whether for physical (hardware) production or for digital (software) production, and specifically those responsible for programs, development projects or test projects, in order to ensure that the objectives of Design organizations are correctly understood, deduced and implemented in the solutions they put in place;

– test managers in charge of quality and system-of-systems testing (at design organization level), as well as test managers in charge of quality and system testing (at design and at client level), applications and predominant software components entering into the composition of systems-of-systems, with the particularity that the so-called "end-to-end" (E2E) tests are not limited to a single application or system, but cover all the systems making up the system-of-systems;

– testers, test analysts and technical test analysts wishing to obtain a more global and general vision of their activities, to understand how to implement their skills and knowledge to further develop their careers;

– anyone wishing to develop their knowledge of testing and their impact on the quality of complex systems and systems-of-systems.

1.2.3. *Organization of this book*

These books are part of a cycle of three books on software testing:

– the first book (*Fundamentals of Software Testing*, ISTE and Wiley, 2012) focuses on the ISTQB Foundation level tester certification and is an aid to obtaining this certification; it was elected third best software testing book of all time by BookAuthority.org;

– this present book on the general aspects of systems-of-systems testing;

– a third book on practical implementation and case studies showing how to implement tests in a system-of-systems, *Advanced Testing of Systems-of-Systems 2: Practical Aspects* (ISTE and Wiley, 2022).

The last two books complement each other and form one. They are independent of the first.

1.3. Examples

We are in contact with and use systems-of-systems of all sizes every day: a car, an orchestra, a control-command system, a satellite telecommunications system, an air traffic control management system, an integrated defense system, a multimodal transport system, a company, all are examples of systems-of-systems. There is no single organizational hierarchy that oversees the development of all the components integrated into these systems-of-systems; some components can be replaced by others from alternative sources.

In this book, we will focus primarily on software-intensive systems. We use them every day: a company uses many applications (payroll, inventory management, accounting, etc.) developed by different companies, but which must work together. This company information system is thus a system-of-systems.

Our means of transportation are also systems-of-systems: the manufacturers (of metros, cars, planes, trains, etc.) are mainly assemblers integrating hardware and software designed by others.

Operating systems – for example, open source – integrating components from various sources are also systems-of-systems. The developments are not carried out under the authority of an organization, and there is frequently integration of components developed by other structures.

The common elements of systems-of-systems – mainly software-intensive systems – are the provision of a service, under defined conditions of use, with expected performance, providing a measurable quality of service. It is important to think "systems" at the level of all processes, from design to delivery to the customer(s) of the finished and operational system-of-systems.

Often, systems-of-systems include, within the same organization, software of various origins. For example, CRM software such as SAP, a Big-Data type data analysis system, vehicle fleet management systems, accounting monitoring or analysis of various origins, load sharing systems (load balancing), etc.

The examples in this book come from the experience of the author during his career. We will therefore have examples in space, military or civil aeronautics, banking systems, insurance and manufacturing.

To fully understand what a system-of-system is in our everyday life, let's take the example of connecting your mobile phone to your vehicle. First of all, we have your vehicle, and the operating system which interacts via a Bluetooth connection with your phone. Then, we have your phone, which has an operating system version that evolves separately from your car; then, we have the version of the software app which provides the services to your phone and is available on a store. Finally, we have the subscription that your car manufacturer provides you with to ensure the connection between your vehicle and your phone. This subscription is certainly supported by a series of mainframes, legacy applications and these must also be accessible via the Web. The information reported by your vehicle will certainly be included in a repository (Big Data, Datalake, etc.) where it can be aggregated and allow maintenance of your vehicle as well as improvement in the maintenance of vehicles of your type. This maintenance information will allow your dealer to warn you if necessary (e.g. failure identified while the vehicle is not at the garage, and need to go to a garage quickly). You can easily identify all the systems that need to communicate correctly so that you – the user – are satisfied with the solution offered (vehicle + mobile + application + subscription + information reported + emergency assistance + vehicle monitoring + preventive or corrective maintenance + … etc.).

1.4. Limitations

This book will focus primarily on systems-of-systems and software-intensive systems, and how to test such systems. The identified elements can be extrapolated to physical systems-of-systems.

As we will focus on testing, the view we will have of systems-of-systems will be that of Test Managers: either the person in charge of testing for the client or for the design organization, or in charge of testing a component, product, or subsystem of a system-of-systems, in order to identify the information to be provided within the framework of a system-of-systems. We will also use this view of the quality of systems and systems-of-systems to propose improvements to the teams in charge of implementation (e.g. software development teams, developers, etc.).

This work is not limited to the aspects of testing – verification and validation – of software systems, but also includes the point of view of those in charge of improving the quality of components – software or hardware – and processes (design, maintenance, continuous improvement, etc.).

As part of this book, we will also discuss the delivery aspects of systems-of-systems in the context of DevOps.

1.5. Why test?

The necessity of testing the design of software, components, products or systems before using or marketing them is evident, known and recognized as useful. The objective of the test can be seen according to a system of five successive phases, as proposed by Beizer (1990):

– testing and debugging are related activities in that it is necessary to test in order to be able to debug;

– the purpose of the test is to show the proper functioning of the software, component, product or system;

– the purpose of the test is to show that the software, component, product or system does not work;

– the objective of the test is not to prove anything, but to reduce the perceived risk of non-operation to an acceptable value;

– the test is not an action; it is a mental discipline resulting in software, components, products or systems having little risk, without too much testing effort.

Each of these five phases represents an evolution of the previous phases and should be integrated by all stakeholders on the project. Any difference in the understanding of "Why we test" will lead to tensions on the strategic choices (e.g. level of investment, prioritization of anomalies and their criticalities, level of urgency, etc.) associated with testing.

A sixth answer to the question "why test?" adds a dimension of improving software quality and testing processes to identify anomalies in products comprising software such as systems-of-systems. This involves analyzing the causes of each failure and implementing processes and procedures to ensure the non-reproducibility of this type of failure. In critical safety areas (e.g. aeronautics), components are added to the systems to keep information on the operating status of the systems in the event of a crash (the famous "black boxes"). The analysis of these components is systematic and makes it possible to propose improvements in procedures or aircraft design, so as to make air travel even more reliable.

Adding such a way of doing things to development methods is what is planned during sprint retrospectives (Agile Scrum methodology) and more generally in feedback activities. This involves objectively studying anomalies or failures and improving processes to ensure that they cannot recur.

1.6. MOA and MOE

When talking about systems-of-systems, it is common (in France) to use the terms client project management (MOA) and designer project management (MOE). These acronyms from cathedral building have been taken up in the world of software engineering. They are 100% French-speaking, and represent two different views of the same things:

– the client project owner (abbreviated MOA) represents the end users that have the need and define the objectives, schedule and budget; MOA is responsible for the needs of the company, of the users and of their customers, of the principals, sponsors or stakeholders, of the business of the company. There usually is only one MOA;

– the designer project manager (abbreviated MOE) represents the person (or company) who designs and controls the production of an element or a set of elements making up the system-of-systems; it is all the production teams, with constraints and objectives often different from those of the company and the principals. There could be multiple MOE.

In a system-of-systems, we therefore must take into account this separation between MOA (client) and MOE (supplier) and therefore the two separate views of each of these major players.

When we deal with systems-of-systems testing, we will speak of "test manager", but these can be assigned to a single test level (e.g. for a software subsystem) or cover several levels (e.g. the manager responsible for testing at the project management level).

1.7. Major challenges

Recent statistics[1] show that only 6% of large IT projects are successful and 52% are outside budget, timeframe or lack all the expected functionalities. The remaining 42% are cancelled before their delivery, becoming losses for the organizations.

We can conclude that the most appropriate development and testing processes should be implemented to minimize, as much as possible, the risks associated with systems-of-systems. When compared to complex systems, Test Managers of systems-of-systems face and must master many challenges.

1.7.1. *Increased complexity*

Systems-of-systems are generally more complex and larger than complex systems developed by a single entity. We must consider:

– interfaces and interoperability of systems with each other, both logical (messages exchanged, formats, coding, etc.) and physical (connectors, protections against EMP, length of connectors, etc.);

– development life cycles of the organizations and their evolutions;

– obsolescence of components of the system-of-systems, as well as their versions and compatibilities;

1 According to https://www.standishgroup.com/sample_research_files/BigBangBoom.pdf.

– integration of simulation and decision support tools, as well as the representativity of these tools with regard to the components they simulate;

– governance and applicable standards – as well as their implementation – for both process and product aspects;

– design architecture and development process frameworks;

– the quality of requirements and specifications, as well as their stability or evolution over time;

– the duration of the design process to develop and integrate all the components, compatibility of these with each other, as well as their level of security and the overall security of the entire system-of-systems;

– organizational complexity resulting from the integration of various organizations (e.g. following takeovers or mergers) or the decision to split the organizations, to call on relocated external subcontracting (offshore) or not;

– the complexity of development cycles stemming from the desire to change the development model, which implies the coexistence of more or less incompatible models with each other for fairly long periods.

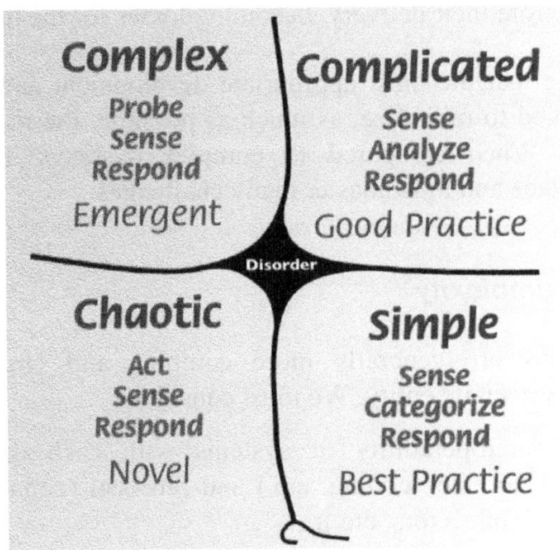

Figure 1.3. *Simple–complicated–complex–chaotic*

We could use the CYNEFIN[2] model (see Figure 1.3, simple–complicated–complex–chaotic) to better understand the aspects of evolution between simple

2 https://www.le-blog-des-leaders.com/cynefin-framework/.

systems (most software developments), complicated systems (e.g. IT systems), complex systems (the majority of systems-of-systems) and chaotic systems, where the number of interactions is such that it is difficult (impossible?) to reproduce and/or simulate all the conditions of execution and operation of the system-of-systems.

To determine if the system is simple, complicated, complex or chaotic, we can focus on the predictability of effects and impacts. We also have the "disorder" state which is the initial position from which we will have to ask ourselves questions to determine which model of system we should turn to.

1.7.1.1. *Simple*

If the causes and effects are well known and predictable, the problem is said to be "simple". The steps can be broken down into feeling, categorizing and then acting. We can look at the applicable "best practices" and select the one(s) that is(are) appropriate, without needing to think too much.

1.7.1.2. *Complicated*

An environment will be said to be "complicated" when the causes and effects are understandable but require a certain expertise to understand them. The domain of practices – including software testing practices – is that of "best practices", known to experts and consultants and making it possible to reach a predefined final target.

1.7.1.3. *Complex*

In the realm of the "complex", the causes and effects are difficult to identify, to understand, to isolate and to define. It seems difficult, if not impossible, to get around the question. We are moving here from the field of "best practices" to that of the emergence of solutions appearing little by little, without an a priori identification of the final target. We are no longer here in a posture of expertise but are entering into a posture of a coach who asks questions, who enlightens through reflections and makes the actors gain understanding.

1.7.1.4. *Chaotic*

In a so-called "chaotic" system, we are unable to distinguish the links between causes and their effects. At this level, the reaction will often be an absence of reaction, like paralysis. When you're in chaos, the only thing you can do is get out of the chaos as quickly as possible, by any means imaginable. Given the exceptional side of what is happening, there are no best practices to apply. You will also not have the time to consult experts who will take a few weeks to analyse in detail what is happening and finally advise you on the right course of action. You will certainly not have the time to do a few harmless experiments to let an original solution emerge. The urgency is to take shelter: the urgency is to act first.

1.7.2. *Significant failure rate*

Most systems-of-systems are large – even very large – projects. Measured in function points (e.g. IFPUG or SNAP), these projects easily exceed 10,000 function points and even reach 100,000 function points. Capers Jones (2018a) tells us that on average these projects have a 31–47% probability of failure. The Chaos Report in 2020 confirms this trend with 19% of projects failing and 50% seriously off budget, off deadline or lacking in quality.

Since the causes of failure add up to one another, it is critical to implement multiple quality improvement techniques throughout the project, from the start of the project. The choice of these techniques should be made based on their measured and demonstrated effectiveness (i.e. not according to the statements or opinions of one or more individual). A principle applicable to QA and testing is "prevention is better than cure". It is better to detect a defect early and avoid introducing it into any deliverable (requirements, codes, test cases, etc.), rather than discovering it late. This principle also applies to tests: reviews and inspections have demonstrated their effectiveness in avoiding the introduction of defects (measured effectiveness greater than 50%), while test suites generally only have an effectiveness of less than 35%. This is the basis of the "shift left" concept which encourages finding defects as early as possible (to the left in the task schedule). This justifies providing stakeholders with information on the level of quality of systems-of-systems from the start of design, as well as measurable information for each of the subsystems that compose them. Implementing metrics and a systematic reporting of measures is therefore necessary to prevent dangerous drifts from appearing and leading the project to failure.

1.7.3. *Limited visibility*

Since systems-of-systems are large projects involving several organizations, it is difficult to have complete and detailed visibility into all the components and their interactions with each other. It will be necessary to use documentation – paper or electronic via tools – to transmit the information. In this type of development, these activities will sometimes be taken over by those in charge of Quality Assurance. Test Managers belong to Quality Assurance, focusing mainly on the execution of tests to verify and validate requirements and needs.

The Test Manager will thus have to:

– analyse information coming from lower levels and related to the level of quality of the subsystems or components which are developed and tested there;

– provide information to higher levels, related to the level of quality of the subsystems or components tested at its level.

Each subdivision level of the system-of-systems must therefore receive the level of information necessary to carry out its activities, and be informed of developments that may impact it. This involves a two-way traceability of information from requirements to test results.

1.7.4. *Multi-sources and complexity*

Can also be considered as systems-of-systems complex information systems comprising numerous software of various origins and natures. For example:

– a production management system based on an ERP (e.g. SAP);

– a customer relationship management system (e.g. SalesForce);

– legacy applications based on multiple systems and technologies;

– applications gathering commercial or other data for statistical purposes (Big Data type);

– applications managing websites, etc.

Applications often come from many external – and internal – sources that have different responsiveness. A management system (ERP or CRM type) can release one or two versions per year with a sequential development cycle, while an Internet application (catalog and Internet sales type) can use a DevOps methodology and deliver every week.

These applications interact and exchange information to create a complete and complex information system. Mastering such a system-of-systems requires having a global functional vision to identify the impacts of a change on the entire information system. We mean here "change" both at the functional level (modification of a management rule, addition of a new commercial offer) and at the technical level (e.g. addition of a new system, modification of the security rules or data transfer, implementation of new tools, etc.).

1.7.5. *Multi-enterprise politics*

Systems-of-systems are developed by different companies with no common development policy. Some of the components used in systems-of-systems may exist before the design of the system-of-systems; other components are developed specifically for the system-of-systems. Lifetime, design mode and component criticality will certainly be different. The responsibilities of the stakeholders as well as the confidentiality of the information will also be different. All these elements should be considered. Let us see them in a little more detail.

1.7.5.1. *Lifetime*

Components – as well as the systems and subsystems – making up the system-of-system have different life spans; it is common for some components to become obsolete or no longer be supported by their manufacturer even before placing the system-of-systems on the market. We could take as an example the number of PCs that still run the XP operating system, within systems-of-systems such as those of defense or large financial organizations, while Microsoft ended support for XP on April 8, 2014, or the number of legacy systems developed in Cobol nearly 40 years ago and still in operation.

1.7.5.2. *Design mode*

Likewise, companies often have different design styles and different testing requirements. Design models vary from sequential development to Agile development, through all the different styles (V, Iterative, Incremental, RUP, Scrum, SAFe, etc.), each with a different level of documentation and verification/validation activities.

The design mode impacts documentation (in terms of volume and evolution) as well as the frequencies of delivery of work products (punctual deliveries or continuous deliveries).

1.7.5.3. *Criticality*

Often, the level of testing of a software component will vary based on the criticality defined by the publisher. A non-critical component – probably tested with less rigor than a critical component – could be introduced later in a system-of-systems of high criticality. For example, we could have, in a critical system, an intrusion notification that is transmitted by the cellular network (GSM) or by the TCP/IP network. Neither of these networks – otherwise very reliable – are reliable enough only to guarantee that the notification will always be transmitted.

1.7.5.4. *Responsibility*

In systems-of-systems, the responsibility for testing at the system-of-system level is the responsibility of a Test Manager that we will call Product Test Manager (TMP). The tests of the various subsystems are delegated to other Test Managers, who report information to the Product Test Manager. The Product Test Manager may therefore have to combine and synthesize the results of the tests carried out on each of the systems making up the system-of-systems.

Industrial organization may involve subcontractors to develop certain systems or components. Each subcontractor must ensure that the deliverables provided are of good quality, that they comply with applicable requirements. Subcontractors sometimes wish to limit the information provided to their client (including the level of tests executed and their results), thus limiting the reporting burden (of measuring, summarizing and processing information). The corollary is that the Product Test Manager is unable to identify tests already carried out and their level of quality.

As the system-of-systems produced will be associated with a brand, it is important for the Product Test Manager to ensure that the quality level of each of the systems that compose it are of good quality.

Example: a system-of-systems produced by the company AAA is made up of many systems developed by other companies, for example, XXX, YYY and ZZZ. If the interaction of the systems produced by the companies XXX and YYY leads to a malfunction of the component produced by ZZZ, it will always be the name of the AAA company which will be mentioned, and not those of the other companies XXX, YYY or ZZZ. It is therefore important for the Project Test Manager of company AAAA to ensure that the products delivered by companies XXX, YYY or ZZZ function correctly within the AAA system-of-systems.

1.7.5.5. *Confidentiality*

Each company has its own processes, techniques and methods which represent its added value, its know-how. In many cases, these processes are confidential and undisclosed. This confidentiality also applies to the testing activities of the products designed.

However, this confidentiality of the processes should be lifted for customers, to allow them to verify the technical relevance of the existing processes. The provision of numerical references and representative statistics, as well as comparisons or references with respect to recognized technical standards, allows an assessment of the relevance of these processes.

1.7.6. *Multiple test levels*

In a system-of-systems, there are many levels of testing. We could have seven test levels (such as for airborne systems):

– testing at the developer level, in a software development environment (see unit test);

– testing the software alone (see section 6.4 of this book), in a separate test environment;

– testing the software integrated with other software in a separate test environment (see integration test and system test within the meaning of ISTQB, system tests within the meaning of TMap-Next);

– testing the software installed on the hardware equipment (see hardware–software integration test);

– testing the equipment in connection with other equipment to form the system, but on a test bench;

– testing the system on the aircraft, on the ground;

– testing the system on the aircraft, in flight (see acceptance test).

If we take another type of system-of-systems – for example, a manufacturing company that markets its products and uses outsourcing – the test levels (here eight levels) may be different:

– testing at the developer level within the development organization (component or unit testing);

– testing of the embedded software, in a separate test environment, within the development organization;

– functional acceptance test of the integrated software components, by the client organization – or its representatives – in the development organization test environment (factory acceptance test or system test as per ISTQB terminology, validation according to TMap-Next terminology);

– acceptance testing of the software integrated with the hardware equipment (by the client organization) in the environment of the development organization (Hardware–Software Integration Testing);

– test of the software installed in the hardware and software environment of acceptance of the customer company including the technical verification of the flows between the various applications of the system-of-systems (integration test and verification of the inter-application flows);

– software testing in an environment representative of production to ensure performance and correct operation for users (acceptance testing);

– testing the software and the system-of-systems on a limited perimeter (pilot phase), but with functional verification of all flows;

– the test of the system-of-systems during a running-in phase (acceptance test).

Each level focuses on certain types of potential defects. Between the various levels, we could have intermediate levels, such as the FAI (First Article Inspection) which check the first equipment – or system – delivered to the assembly line. Each level has its own requirements, development and test methods and environments, development and test teams.

The number of test levels, and separate test teams, increases the risks:

– of redundant tests, already executed at one test level and re-executed at another level, which impacts test efficiency;

– of the reduced efficiency of the test activities, because the principle is to remain focused on the objectives of a level without re-executing the tests of a lower level. This can be mitigated by analyzing the feedback of the level test coverage information to higher levels.

For business information systems and systems-of-systems, it is also necessary to consider other levels of acceptance of applications or systems:

– Does the software integrate and communicate well with other software? A systems integration testing level may be required.

– Has the distribution and production of the application been checked? Is it easy, is it possible to rollback – without losses – in the event of problems?

– Have the training activities been carried out to allow optimal application use?

– Are metrics and measurements of user satisfaction with the application considered to ensure customer satisfaction?

One of the usual challenges for a Test Manager and their test team is to be effective and efficient, that is, not to perform unnecessary tasks. A task that does not add value should be avoided. When certain tests are carried out by a development team (outsourced or not) and the same tests are carried out at another level, we lose efficiency. It is therefore essential to have good communication between the teams and identification of what is being done at each level of testing.

Defect detection and remediation costs increase over time; teams should be directed to perform testing as soon as possible to limit these costs. Each design activity should be followed by an activity that will focus on identifying defects that might have been introduced in that design activity. Phase confinement can be measured easily by an analysis of the phases of introduction and detection of defects. The ODC technique is based on this principle to identify the processes to be improved (see also section 13.3.1.3 of volume 2).

In subcontracted and/or outsourced development and testing, tests should be carried out at each level, both by the subcontractor's design team and by the client teams. The tests executed by the design team are often more detailed and more technical than the tests executed by the customer team (which rather executes functional tests). The lack of feedback from the design team to the customer team prevents the implementation of an optimized multi-level test strategy.

1.7.7. *Contract follow-up, measures, reporting and penalties*

The principle of a system-of-system is that the development of components, products and systems and their testing are not done under the authority of the same management, but under different organizations. It will be necessary to define the contractual relations between the partners and between the systems of the system-of-systems.

Contracts should include synchronization points or information feedback allowing the main contractor – the project owner – to have a global and sufficiently detailed view of the progress of the realization of the system-of-systems. The feedback information is not only information on deadlines or costs, but should also cover the quality of the components, their performance, reliability level and relevance regarding the objectives of the system-of-systems. As mentioned earlier, each system may require multiple levels of testing. The results of each of these test levels must be compiled and integrated in order to have a global view of the project.

1.7.7.1. *SLA and penalties*

Each development and each test level can be the subject of a specific contract, including objectives in terms of service quality (SLA; Service Level Agreement) or product quality, metrics and progress indicators.

The definition of the metrics, and the measures necessary to ensure the level of quality of each product, must be included in each contract, as well as the frequency of measurement and the reporting method. Non-compliance with or failure to achieve these SLAs may result in contractual penalties.

It must be realized that penalties are double-edged swords: if the level of penalty reaches a threshold that the subcontractor cannot bear, the latter may decide to stop the contract and wait for a judicial resolution of the dispute. Even if the litigation ends with a condemnation of the subcontractor, this does not solve the problem at the system-of-systems level. The principal will be forced to find another subcontractor to replace the defaulting one, and there is no indication that this new subcontractor will accept binding penalty clauses that do not suit them. We have had the experience of subcontractors who, having reached the maximum level of penalties provided for by the contract, decided to suspend performance of the contract until the penalties were waived.

1.7.7.2. *Measures, metrics and reporting*

The metrics and measures must be adapted to each level of the system-of-systems, to each level of progress of the project, and according to the software development life cycles (SDLC) selected for the realization.

The measures must cover the product, its level of quality, as well as the processes – production and testing – and their level of efficiency. In terms of reporting, information must be reported with sufficient frequency to make relevant decisions and anticipate problems.

To ensure that the information is unbiased, it is important to measure the same data in several ways or at least to ensure its relevance. For example, considering that a software is of good quality if no defect is found during a test period is only valid if during this period the tests have been carried out on the application. That is, the application has been delivered and is working, the test cases and their test data have been executed correctly, the expected test results have been obtained, the desired level of requirements and code coverage has been achieved, manual testing has been performed by competent testers, etc.

Frequently, outsourcing teams are tempted to hide negative information by disguising it or glossing over it, hoping that delays can be made up for later. Therefore, it is essential to carefully follow the reported measurements and investigate inconsistencies in the ratios.

1.7.8. *Integration and test environments*

In a system-of-systems, exchanges between systems and between applications are important. These exchanges are tested during integration tests. In the absence of components, it may be necessary to design mock objects (stubs, emulators or simulators) that will generate the expected messages.

These objects may vary depending on the test levels. They will have to evolve according to the evolution of the specifications of the components they replace, and may have to be scrapped when the components they replace are made available.

The main concerns associated with these objects are:

– their (in)ability to generate all possible combinations of flows, whether planned combinations or unplanned combinations. The combination of message types, how they are formed, their sequencing, etc., is important and may require testing over a period of time;

– evolution of component versions and the ability of components to communicate with each other, which requires configuration and version management of the components in each environment, with an evaluation of their quality via tests;

– the increase in exchanges between components, applications and systems to transfer more information, which may lead to saturation and bandwidth reduction or even latency increase.

1.7.9. *Availability of components*

The design of a system-of-system integrates existing components and anticipates the availability of components under development.

Reusing systems or components that have already proven themselves makes it possible to limit procurement costs and lead times. There is no guarantee that the systems or components used will be maintained throughout the life of the system-of-systems being designed. For example, if we integrate a PC card running under a Windows 7 OS into a rack, there is no guarantee that this card or this OS will be supported by its manufacturer or its publisher throughout the lifetime of the system-of-systems. We could therefore be forced to replace the card with another card in the future (will it keep the same physical and electronic characteristics and the identical electronic interfaces?), or to replace the OS with another (will it have the same behavior as the initial one, the same interfaces, the same messages, etc.?).

Similarly, the design of a system-of-systems may consider the inclusion of systems that are under development and will only be ready in the future. These components will provide services and interfaces will provide the connection to these services. As long as these components are not available, it will be necessary to replace them with plugs, or dummy components simulating the expected components. Several levels of integration tests will be necessary to ensure the perfect integration of these new components into the system-of-systems.

1.7.10. *Combination and coverage*

The definition of requirements and user stories, in fact the definition of needs, makes it possible to describe a “normal” behavior that it is necessary to test to ensure that it is indeed present. On the other hand, it is behaviors that are not “normal”, incongruous or very exceptional combinations that can generate unexpected events with catastrophic consequences, the “Black Swan” according to Taleb (2008).

In a system-of-systems, modeling usually defines the nominal behavior. This normal operation will be checked by the designers and validated by the testers and the users (e.g. Product Owners, users, etc.). On the other hand, anything that is NOT modeled, but which may have an impact on the model, should also be considered. As these exceptional behaviors have not been modeled, because they are not specified, they will never be found in the user stories or in the requirements, and the designers will be right to answer: “this behavior was never requested in the contractual documents!”

To ensure an adequate level of quality – especially for events that can have a critical impact – testers should test all these combinations which takes a lot of time. It is necessary to understand that a system-of-systems, or a complex system, is composed of many systems – each of which can be modeled – and that the combinations of “normal” or modeled events are very few when compared to non-normal (exceptional), unplanned and therefore unmodeled events.

The analysis of air disasters shows that the root causes of these disasters are very often combinations of rare exceptional cases.

Identification of all combinations and their coverage by tests, including in the cases of exceptional combinations, is therefore critical.

1.7.11. *Data quality*

In systems-of-systems, as in complex systems, the impact of poor quality data is significant. Data structures can evolve, data can be obsolete, or no longer correspond to the types of data currently expected, or even reflect old data typologies that are no longer being used.

Prior to the execution of the test activities of a level, it is very important to ensure the presence and the quality of the test data that will be used. This will allow:

– focusing tests on existing processes;

– ensuring that data fit expectations (accept valid data, reject invalid data).

During test activities, beyond processing the expected data, it will be necessary to ensure the correct processing of data in an old format or different from the expected format: should this data be rejected, should it be logged for verification and/or correction purposes, or will they have to be processed, for a given period, as correct data?

During system integration testing, it will be necessary to ensure that system data is correctly synchronized, both during the initialization of environments and throughout the system integration test campaign.

1.7.12. *Flows, pivots and data conversions*

Systems-of-systems often use ETL (Extract, Transform, Load) and/or EAI (Enterprise Application Integration) to federate applications around a core – middleware – managing stream data conversions and their transfer from one application to one or more others. Implementation of such a solution involves the creation of a repository of all data (all business objects) and all flows (transmissions and reception) used by system-of-systems applications.

The challenges brought by these middlewares come from their mode of operation:

– some (ETL) mainly have a batch-oriented operation and massive data processing, unidirectional, sometimes used in a BI (Business Intelligence) framework;

– others (EAI) have an event-driven and bidirectional mode of operation, allowing, for example, the conversion and use of the same data by several applications of the system-of-systems;

– still others (ESB for Enterprise Service Bus) allow – like EAI – us to transfer data between applications, but on the basis of open protocols (e.g. XML for the description of messages and web services for data exchange).

Both use conversion tables and allow the design of service-oriented systems where business repositories can share data. These conversion tables as well as the middleware itself are single points of failure that will need to be secured and controlled.

In the context of tests, as in the context of evolutions and transitions – see section 1.7.13 – it will be necessary to define specific management rules to separate test flows from production flows, and to deal with aspects of different versions of the same stream.

1.7.13. *Evolution and transition*

Systems-of-systems are intended to operate for many years. They will therefore have to evolve. The size and number of impacted systems in a system-of-systems solution requires unsynchronized evolution. Let us take a chain of stores as an example: given the impossibility of changing all the stores simultaneously, some stores will remain in an old version, while others will have already evolved to the new version. The central systems-of-systems will therefore need to process data in both the old format and the new data format.

If we take the interaction between your mobile phone and your vehicle as an example, there are many components that evolve separately: the applications on your mobile and the operating system of your mobile, the mobile operator with which you have contracted for your connection services (3G, 4G, 5G, etc.) as well as all its management systems, you have the operating system of your vehicle, that of the various visualization interfaces (graphics tablets, WiFi transmitters/receivers or Bluetooth), the communication systems between your vehicle and the manufacturer's IT system, the subscription that you have taken out with the manufacturer, exchanges between the manufacturer's IT system and those of garages and/or dealers of the brand for the processing of maintaining your vehicle, managing and updating your GPS maps, etc. Of course you also have the many integrated circuits (management of brakes, tire pressure, fuel injection into the engine, management of the remaining fuel and calculation of your autonomy according to your driving mode, diagnostic systems, trajectory control, etc.) which will provide information which should be displayed on your vehicle's control screens.

Each of these systems evolves without synchronization with the others, and the objective is that the service continues to be provided throughout the chain, without negative impact. Admittedly, in the case of a vehicle, the system-of-systems is not specifically critical, although certain components are (power steering, brakes if they do not trigger, air bags if they trigger unexpectedly, injection if it stops under full acceleration, etc.).

1.7.14. *History and historization*

Systems-of-systems are often required to keep track of the exchanges that have taken place between systems, to be able to use this information for analysis or evidence purposes. Data, for example, those passing through an EAI or an ETL (see section 1.7.12), must be stored or logged for periods of up to several years. This leads to both problems of data storage volume and of securing this historical data to meet regulatory requirements (e.g. GDPR) and access to this data.

It is tempting for the tester to use this historical data, but it is then necessary to anonymize it to respect the applicable security and confidentiality rules. Another challenge with using historical data is that – in very many cases – (1) the data is incomplete (see volume 2, section 15.2.4) and (2) this data represents various instances of the same equivalence partitions. The use of such data does not bring any added value for the detection of faults in these partitions. If these elements are considered, this data can be useful for measures of performance, robustness and reliability.

1.7.15. *Impostors*

Many claim to be recognized experts in their profession or to be companies that have reached very high levels of maturity (e.g. CMMI level 5, TMMI level 5, etc.), thus justifying confidence in their abilities. The analysis of their processes and practices reveals a vision that is the least "optimistic" of their skills. We could illustrate this by counting the number of companies rated CMMi level 5 in India (134), the USA (52) and France (none)[3]. As far as individuals are concerned, it takes at least 10,000 hours of practice, or more than five years, before reaching a truly advanced level. The "expert" level requires much more practice in different fields and environments before being able to claim that title.

A high level of maturity matters when awarding subcontracts: a client will select – at equivalent cost – a supplier claiming to have a high level of maturity. This use of maturity information for marketing purposes can mislead buyers.

When it comes to software testing, just look at the number of companies highlighting software testing skills and the low number of software that are delivered on time and on cost: 6%. Are testing skills overrated by companies, or are we dealing with purely marketing information?

Many testers claim to be "expert in testing" with less than 5 or 10 years of expertise, and even though the number of French-language publications on software testing is small, seldom have more than one book on the subject of testing. Globally, the ISTQB announces that out of a base of more than 440,000 testers only 0.7% have a full advanced level and 9.2% have one of the three advanced level certifications. This compares to 90.1% with only the foundation certification which

3 According to https://sas.cmmiinstitute.com/pars/pars.aspx.

can be obtained in three days of training. It is therefore important to understand the differences between the levels of certifications and the significant gap between an advanced level and a foundation level. If there is one suggestion we can make, it is to focus on testers with at least one – and if possible all three – advanced ISTQB certifications.

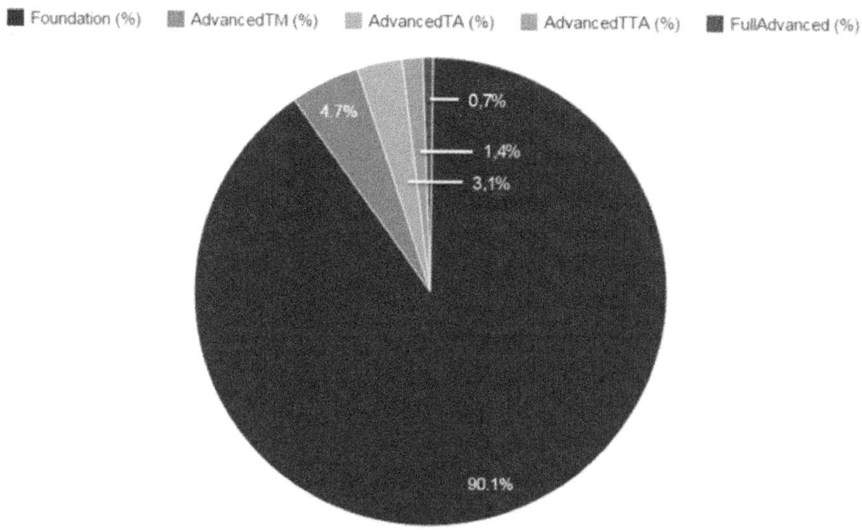

Figure 1.4. *ISTQB foundation versus advanced testers. For a color version of this figure, see www.iste.co.uk/homès/systems1.zip*

Acquiring a certification does not in any way guarantee human or relational skills of individuals or organizations, especially if this certificate is used solely for commercial or employability purposes, and is not supported by significant and demonstrable experience, in the field of the test, the project or the system-of-system to be tested.

Acting like a foundation level certification has the same value as an advanced or expert level certification misleads both the client and the co-contractors on the level of competence of the individuals and the quality of the processes used.

A systematic verification of references and exchanges with the people or projects cited as references is one way of limiting the impact of such types of people. This is also valid in the case of subcontracting or co-contracting companies: it is the individuals who have enabled the success of past projects. There can be no assurance that the individuals assigned to your system or your system-of-systems have the knowledge or abilities of those who have managed the referenced projects, or have the skills necessary to adapt to your projects. Even though reference checking can be considered as a lack of trust, it goes hand in hand with testing activities where "Trust but verify" is the basis.

2

Software Development Life Cycle

The systems and software that make up a system-of-systems are developed by different organizations, using different development cycles. Hardware components are often developed in sequential or incremental design cycles. Software can be developed with other development cycles.

There are various software development cycles, called SDLC (for Software Development Life Cycle). Their purpose is to deliver good quality software in compliance with requirements, deadlines and costs.

Please do not confuse "development cycle" and software "life cycle". The development cycle ends when the software is delivered, whereas the life cycle includes, in addition to the development cycle, the maintenance and disposal of this software, or even its replacement by another software.

Whatever development cycle is selected, it must meet partially contradictory objectives:

– it must allow delivery on time and within budget of a product that is efficient and of sufficient (known) quality;

– the product must be documented, maintained and supported for several years;

– since the system-of-systems is associated with the company that markets it, it is up to the latter to ensure its level of quality, including the quality of each of the components.

A fast delivery implies a reduction of the time of design and realization which can lead to a reduction of the verification and validation activities (i.e. testing) of the components, and therefore to a drop in quality. Increasing profitability is often understood as not performing one or more activities. In fact, profitability is a balance between reducing the number of things to do and reducing the quality of the product. A reduction in product quality will increase the number of things to do, such as finding faults, correcting them, and retesting them and then re-delivering.

> Better to do it right the first time than to do it, then detect the defect, then fix that defect and then retest to make sure the defect is fixed and has not introduced regressions elsewhere.

In a system-of-systems environment, given the presence of hardware components, sequential development cycles and incremental design will often be used: a first delivery including the minimum functionalities (MVP for Minimum Viable Product) followed by improvements or evolutions over the years (e.g. evolution of versions of equipment of the same type).

> **Increments or Iterations?**
>
> In iterative development, each iteration brings a greater level of detail for the whole developed component. System functionality may not be fully defined at the start of the project.
>
> In incremental development, each increment adds and integrates new functionality to the system being developed. However, the system architecture is determined at the start of the project.

Any modification of the business needs – initially recorded in the definition of the requirements – will force an analysis of the side effects by re-executing all or part of the activities previously carried out. Impact on deadlines and costs is therefore very significant. Note that – in the case of systems-of-systems – the original requirements will certainly evolve over the years and releases.

2.1. Sequential development cycles

Sequential development cycles start from an initial concept (a definition of system-of-systems goals or objectives) that is refined and then realized through a series of sequential operations, although some may be concurrent. This way of developing is easy to understand and is practiced every day in our existence, for example, for the construction of housing where we start with the plans, then we build the foundations, then the floors and the roof, ending with the interior fittings. Some systems-of-systems lend themselves very well to sequential developments, such as the construction of a bridge or a tunnel: as long as the work is not finished, the service is unusable.

In a system-of-systems, there is generally a general design (defined by the client organization) which determines the overall architecture of the components and their interactions. Then, each system is developed and tested by the suppliers, then the systems are then integrated – by the client organization – into the system-of-systems, which is tested from start to finish (end to end) under the responsibility of the client.

Sequential cycles are composed of waterfall cycle and V-cycle.

2.1.1. *Waterfall*

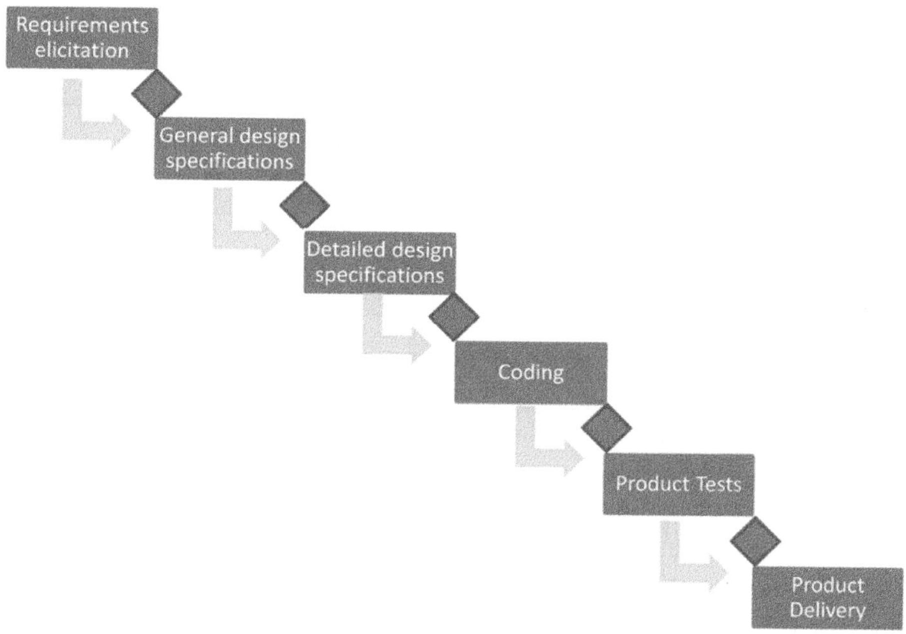

Figure 2.1. *Waterfall*

Waterfall cycle is a sequential representation of activities, where each activity is fully completed before starting the next. The activities are separated from each other by milestones (represented here by diamonds) which ensure that the objectives (exit criteria) of the past activity are achieved and that the launch criteria of the following activities are satisfied.

Each activity develops the concept based on the deliverables generated by the previous activities and considers these input data as correct. Any modification of the requirements and re-executing one or more activities already completed will impact the deadlines and the costs.

This development cycle is frequently used when there is design or integration of hardware components, without prototyping. This development cycle is used, among other things, in the building where plans are designed and built based on these plans, starting with the foundations and ending with the interior layout.

2.1.2. *V-cycle*

In a system-of-systems, the V-cycle is represented as follows:

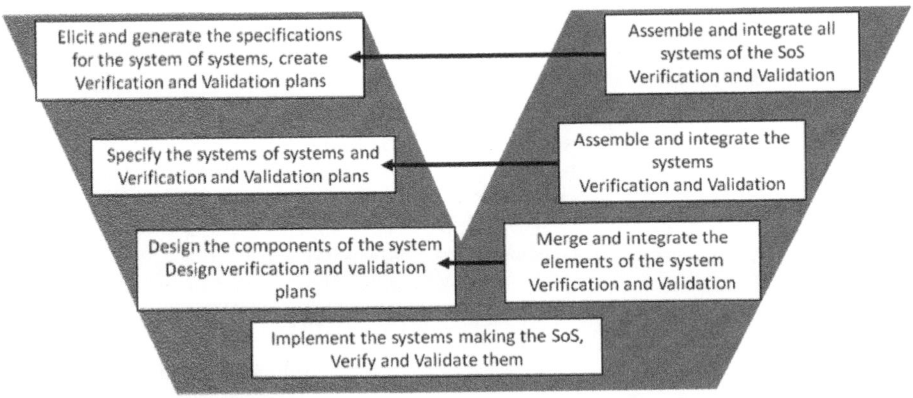

Figure 2.2. *Systems-of-systems V-cycle*

The specification activities (left side) are aligned with the actions on the right-hand side which involve assembling the elements of the system-of-systems and then their verification and validation.

The V-cycle (initially proposed by Barry Boehm in 1985) puts the design activities in front of the test activities relating to the same granularity and defined from the same design documents. The design activities of the components of a level can begin before the formal end of the activity which precedes them. The development of verification and validation activities can be carried out at the same time as the component design activities.

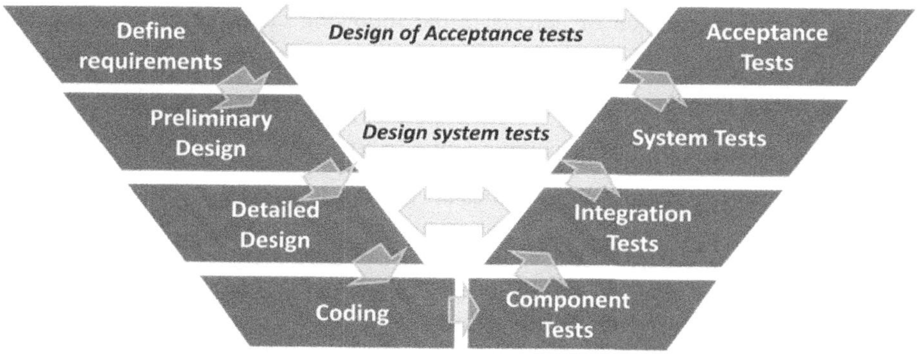

Figure 2.3. *Standard V-cycle*

The representation of the V-cycle, as mentioned in Figure 2.3, suggests an equivalence of the descending and ascending phases, therefore an equivalence of duration. This equivalence is somewhat inappropriate and should instead be represented as Figure 2.4. The terminology we should use would be “a reclining L”.

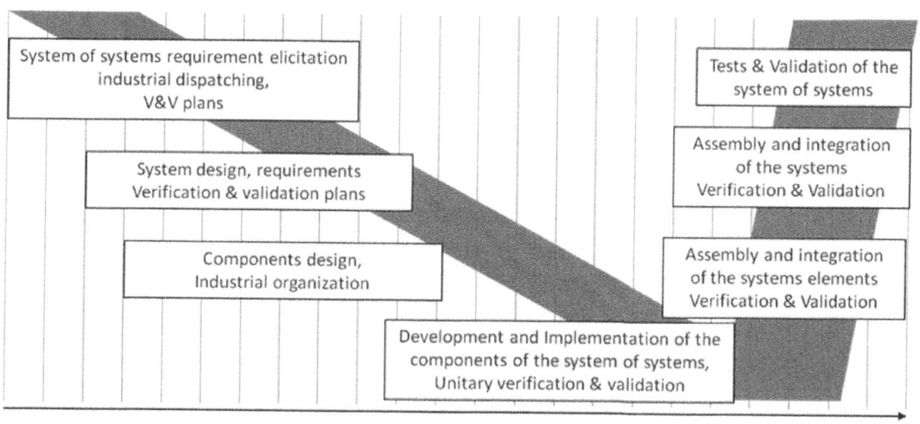

Figure 2.4. *Reclining L cycle*

We see that the durations of the design, analysis and conception phases are longer than the integration and testing phases. This is one of the reasons for involving testing activities (verification by analysis, reviews, test design, etc.) as early as possible, well before the start of the upstream phase. This allows us to:

– avoid the introduction of defects in the following elements by checking the input elements;

– ensure that the elements are complementary and correctly designed;

– anticipate future problems by their early identification;

– propose non-redundant test activities at each test level.

Returning to our sequential cycle, the difficulty of integrating requirements changes during the project remains present. All changes in requirements will require modifications (of architecture, functionalities, etc.), involving an assessment of their organizational and technical impacts, in terms of costs, deadlines and, for example, performance.

In a system-of-systems, the high-level V-cycles can be adapted so that the lower levels, dealing with software, can be different (e.g. Agile, Scrum or SAFe).

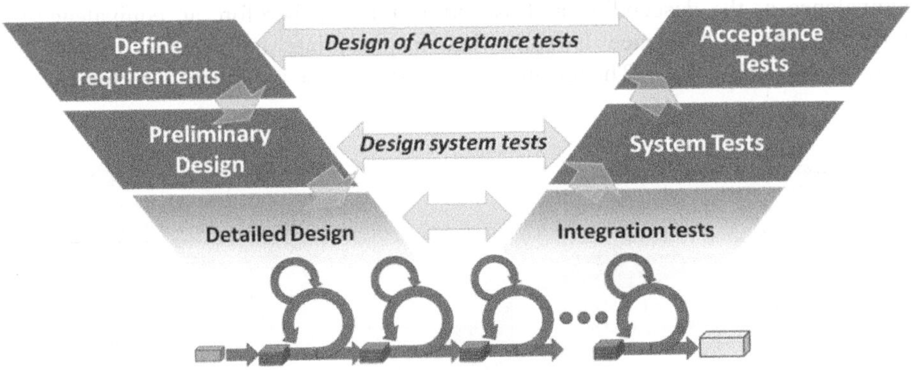

Figure 2.5. *Hybrid V-cycle and Agile*

2.1.3. *Spiral and prototyping*

The spiral development cycle (also proposed by Boehm) includes iterations that result in prototypes. These iterations identify risks, validate operational concepts and requirements (general and detailed) and remove development uncertainties.

When the architecture and the requirements are defined with a sufficient level of detail – here after the three iterations planned on the diagram (Figure 2.6) – a sequential development is implemented, with all the necessary specification and test documents.

The primary interest of the spiral cycle is the emphasis on risk analysis as the design evolves. The main disadvantage of the spiral cycle is that we often tend to develop the final system from the prototype and not to use the prototype as a reference. The result is that the prototype, often developed "in a hurry", is used as the basis of the final product, without cleaning up or commenting on the code of the prototype.

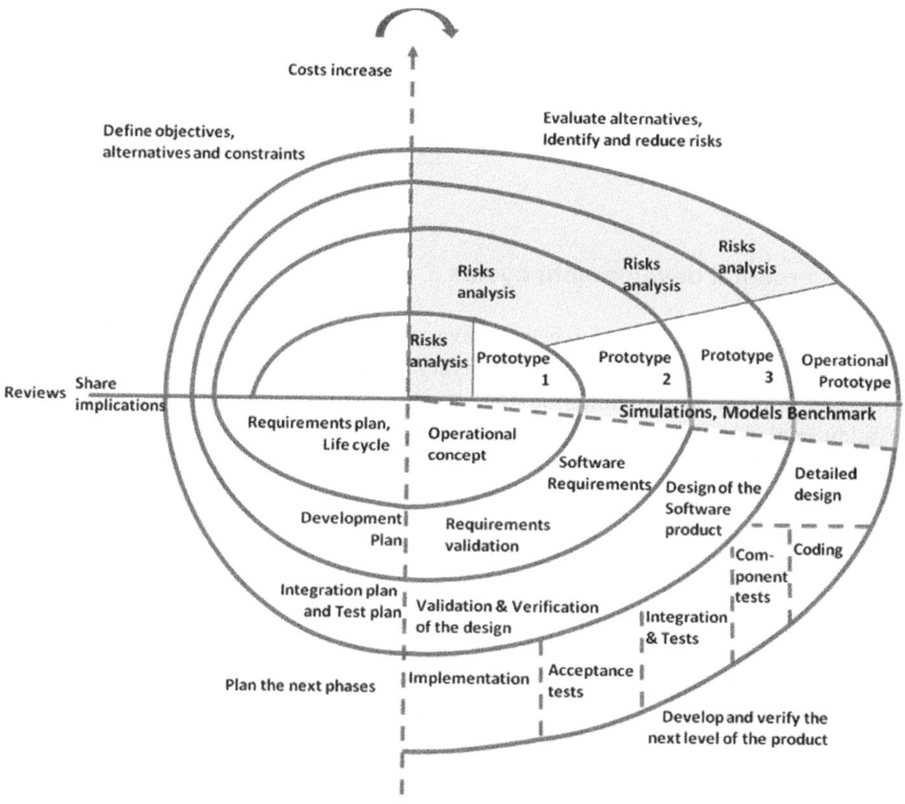

Figure 2.6. *Spiral development cycle*

2.1.4. *Challenges of sequential developments*

Advantages	Disadvantages
Easy to understand	Changing needs and requirements throughout the project, resulting in last-minute changes
Ease of definition of decomposition and subcontracting	Concerns in the completeness of the requirements, and in their definitions
Easy adaptation to systems-of-systems	Development activities tend to take longer while the delivery date remains fixed. This results in a reduction in test times, and therefore a reduction in tests.
	Low defect tolerance; defects are often found late, causing significant delays and costs

2.2. Incremental development cycles

With systems-of-systems, we can have so-called "incremental" developments. These developments consist of a first delivery of the target system-of-systems with a minimum number of functionalities or services, followed by subsequent deliveries of additional functionalities or services. We can, for example, think of the IOC (Initial Operational Configuration) and FOC (Full Operational Configuration) deliveries of aircraft, where the IOC represents the first deliveries, and FOC represents later deliveries having the capacity to integrate corrections of bugs or new features.

Incremental development principle is to split the development into several deliveries, each delivery representing a major evolution compared to the previous version, the addition of new functionalities. We could think of a series of V-cycles that follow each other. Incremental development assumes a complete initial vision of the product to be delivered.

The principle of continuous delivery (CD) proposed by Agile developments (e.g. DevOps), although incremental in spirit, offers too frequent increments – for example, one per month or one per week – involving a completely different validation and delivery.

Unlike sequential developments, in the event of a change in business needs, the impact analysis is carried out without repeating all the initial activities, but by focusing on an analysis in relation to the existing one. The impact on lead times and costs is then – theoretically – lower. However, this implies that the elements on which the impact analysis will be based (documentation, analysis, specifications, architecture, data dictionaries, etc.) have been exhaustively described and correctly maintained and that impact analysis focuses on all potential regressions, both close and distant.

2.2.1. *Challenges of incremental development*

Incremental development faces several challenges:

– impacts of current increment on previous iterations (software configuration management):

- monitoring of versions already delivered and obligation – or not – to update the versions,
- ability of different versions of software to interface with each other;

– impact of delays of one increment on the following increments, cascading across multiple tasks and environments;

– the regression test load, the obligation to maintain the various test environments for each of the levels.

Advantages	Disadvantages
Allows definition of a delivery schedule, taking into account business needs	Delays accumulate
	Need to test regressions of previous versions

2.3. Agile development cycles

Agile development cycles position themselves in opposition to the waterfall cycle, with the aim to provide visibility to users and responsiveness to changing business needs.

Agile development cycles are iterative and incremental cycles, the objective of which is to pragmatically and incrementally develop a system. They differ from more traditional development cycles by taking changes into account and reducing formalism. In general, these "agile" methods respect the four fundamental values and the 12 principles defined in the Agile Manifesto.

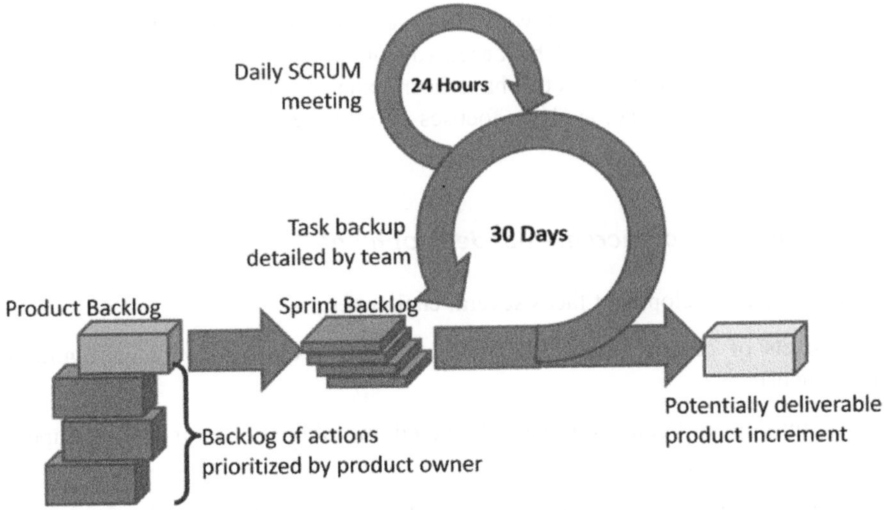

Figure 2.7. *Cycle Agile – Scrum*

Various "Agile" methods exist; the best known are eXtreme Programming and Scrum. These methods, focused on small projects, can be integrated into more complex frameworks such as NEXUS or SAFe. There is sometimes a mix between the methods and their reporting as in the case of KANBAN.

In an Agile development cycle, the members of the development team decide which features to develop during each iteration. At the end of the iteration, the developed functionalities are potentially deliverable.

By definition, an Agile development cycle is incremental. The availability of features over time cannot be anticipated, given the freedom left to the production teams. It is therefore difficult to assure customers, at the start of the project, of the delivery date of all the functionalities with the desired level of quality. However, this development cycle is designed to easily manage changes in customer needs.

2.3.1. *Agile Manifesto*

The Agile Manifesto [AGILE2001] proposes four values and 12 principles. Below are these values and principles along with some thoughts on their suitability for use in systems-of-systems.

Value #1: people and their interactions are more valuable than processes and tools. As much as this value is interesting for components and software products of reduced size, when the duration of design and use of a system-of-systems is important, the need to have a repository, tools and processes – even common standards – and shared documentation becomes important.

Value #2: software that works rather than exhaustive documentation. Can this vision really apply to systems-of-systems where the life span of software is long and where the complexity of interactions is such that a single person cannot master everything?

Value #3: collaboration with customers rather than contract negotiations. This vision seems fragmented and difficult to implement on a system-of-systems which is based on interactions between different organizations for which the contractual aspects are very important.

Value #4: adapting to change more than following a plan. The follow-up of a plan is always subject to hazards, and change will certainly take place. The absence of a plan or its frequent modification will have an impact on the parties in charge of the other components of the system-of-systems, and therefore on the ability to achieve the desired system-of-systems.

– Principle #1. The highest priority is to satisfy the customer by quickly and regularly delivering features with high added value. This implies that the value of a feature – as assessed by the client – is accepted by all stakeholders and that the other components and their dependencies are available and with an adequate level of quality.

– Principle #2. Welcome changing needs positively, even late in the project. Agile processes harness change to give the customer a competitive advantage. The impact of changes varies depending on whether they take place before, during or after the sprint where the part of the software to which the change relates is developed. If before, an impact analysis must be carried out – at the time of development of the change – to define the scope and the effort associated with the change; if during or after, the change will not be implemented immediately, but considered a new requirement and added to the project backlog.

– Principle #3. Frequently deliver operational software with cycles of a few weeks to a few months and a preference for the shortest.

This principle is interesting, insofar as the delivery processes make it possible to ensure the provision of real added value to customers. If the distribution of the software depends on marketing campaigns, requires training or certification efforts, the durations may have to be extended.

– Principle #4. Users or their representatives and developers must work together daily throughout the project. This is difficult to implement when talking about system-of-systems users versus component developers of a subsystem or component, partly because the workload of user representatives is performed in addition to their current workload.

– Principle #5. Carry out projects with motivated people. Provide them with the environment and support they need and trust them to achieve the goals set.

– Principle #6. The easiest and most effective method of conveying information to and within the development team is face-to-face dialogue. This principle is valid for any short-term interaction, but unacceptable for projects lasting several months or several years. In such a case, the traceability of decisions and achievements is necessary and cannot be limited to oral actions.

– Principle #7. Working software is the main measure of progress. This principle can be extended to components, subsystems and systems, but implies that the operational capacity of the component (or software) can be properly verified and validated.

– Principle #8. Agile processes encourage a sustainable pace of development. Together, sponsors, developers and users should be able to maintain a steady pace indefinitely. This principle promotes a smoothing of the workload without the bumps associated with increased workloads just before deliveries.

– Principle #9. Continuous attention leads to technical excellence, and good design enhances agility. This principle implies that design excellence is evolutionary. In cases of technological or technical leaps, this assumption is incorrect.

– Principle #10. Simplicity – the art of minimizing the amount of unnecessary work – is key. This principle is often difficult to implement on systems-of-systems, given their complexity.

– Principle #11. The best architectures, specifications and designs emerge from self-organizing teams. This principle affirms that the optimal architectures and designs emerge from a consensual vision by the implementation team. This depends on the skill level of the team members and would tend to limit Agile in experienced developer teams. This statement – about the emergence of an optimal design – does not correspond to the author's experience.

– Principle #12. At regular intervals, the team thinks about ways to become more efficient, then adjusts and modifies its behavior accordingly. This principle of periodic process improvement is a very good thing and is not limited to Agile processes. It is also necessary to train team members in the planned improvements, and probably measure the effectiveness and efficiency of these improvements.

2.3.2. *eXtreme Programming*

Like the Agile Manifesto, eXtreme Programming has a set of values and practices. According to Ron Jeffries "eXtreme Programming is a software development discipline based on the values of simplicity, communication, feedback, courage and respect. It works by bringing the whole team together in the presence of simple practices, with enough feedback to allow the team to see where they are and adapt the practices to their unique situation". This definition is similar to that of the Agile Manifesto. The differences come from the four values and the 12 principles that make them up:

– *communication*, which involves the presence of a customer representative on site, the implementation of pair programming and coding standards;

– *simplicity*, which includes metaphors to remember naming conventions, simple design limited to the problem at hand, refactoring to remove duplicate or unnecessarily complex code;

– *feedback*, which includes unit testing which must be written by designers, continuous integration which must ensure unit tests are 100% working, and small releases with few new features to that time;

– *courage*, including the planning as well as the negotiation between client and developers of the workload, the ownership of the code shared by the whole team and the establishment of a maintainable velocity over time;

– *respect* needed towards and between team members, to improve communication, provide and accept feedback – including criticism – and work together to identify simple designs and solutions.

These principles are interesting, but their implementation is not always efficient:

– presence of a client in the team implies that this individual – by default a person with knowledge – becomes unavailable to the client company at least for part of the time of the project;

– pair programming is interesting, but the advantage of two pairs of eyes is offset by the need to pay two salaries;

– use of coding standards is a good solution, but can be a brake on the creativity of designers;

– metaphors to remember naming conventions is a nice solution, but a naming checklist may be easier to implement;

– a simple design limited to the problem to be solved may be profitable, but does not make it possible to anticipate future needs and provide useful improvements such as designing reusable routines from the design stage (instead of designing them during a refactoring);

– refactoring is a necessary practice and often improves code quality by removing duplications, but refactoring is not always the most efficient solution: to reverse what was done is less efficient than doing it right from the start. Moreover, it is necessary to regression test the whole software to ensure refactoring did not generate inadvertent side effects;

– feedback from unit testing and integration testing is of course necessary, but are designers going to test something they didn't design? If we stick to the simple design, limited to the problem at hand, then no. Moreover, unit tests are not sufficient in the context of systems-of-systems;

– limited and small deliveries make it easier to identify side effects and regressions, provided that exhaustive regression testing campaigns are carried out. It is a cumbersome solution and not very efficient;

– planning through cooperation between clients and designers can only work if both parties remain reasonable in their demands. However, human beings always want to have more for the same price (or less) and the complexity of a requested functionality is often discovered as development progresses;

– collective code ownership by the team and the need for each person on the team to be able to modify any part of the system implies that everyone on the team has a similar level of skill, even for people new to the team;

– a sustainable velocity over time is interesting and somewhat contradictory to the term "sprint" in agile methodologies.

We see that, as with Scrum, XP is applicable for smallish projects but probably less easy to implement in systems-of-systems.

2.3.3. *Challenges of iterative cycles*

Iterative development methods aim to define the product by refining it over several iterations. Unlike incremental development methods which required defining the complete product from the outset and breaking this product into increments, iterative approaches do not define – from the start – the final product: they propose a vision of the product, which is refined and detailed with each iteration. The final product may therefore not correspond to the product initially envisaged.

Agile methods – Scrum, XP, etc. – were initially designed for small projects: the size of the teams proposed in these methodologies ranges from six to nine people. This allows rapid exchanges between developers and testers (DEV), on the one hand, and business representatives (PO), on the other. Systems-of-systems require the integration of many teams, each dealing with a specific subject. Methodologies for integrating agile developments at scale have been proposed such as Nexus or SAFe. These methodologies involve several interesting challenges:

– The need to separate developments and systems between multiple teams who must share the same objectives, interact with each other, while potentially being geographically dispersed. This implies reliable communications and these communication supports become a potential point of weakness.

– The obligation to keep the developments within the duration of the sprints (e.g. three to four weeks) requires a detailed breakdown of the activities so that each can be carried out within the framework of a single sprint. This implies that these User Story refinements should take place in previous sprints and limit the size of developments to something that can be coded and tested over the duration of the sprint. We will therefore have activities that will overlap, possibly over three sprints. Similarly, cleaning sprints – refactoring – will be necessary every three iterations in order to simplify, reorganize, document the developed code.

– The need to test the impacts of developments in the current iteration on previous iterations and on other components, which will likely require multiple concurrent test environments, including emulators or simulators to replace hardware components not yet available. Regressions or side effects generated by a modification of a component on the other components must also be processed. Theoretically, any modification of a component or its environment can have an impact (a side effect) and should involve the re-execution of all the tests. This is obviously impossible to envisage in a reasonable time and at a reasonable cost. It will be necessary to ensure a management of the regression tests applicable to the chosen development model.

– Sharing and/or assignment of responsibility. Let us take a few examples: (a) in the case of interface concerns between two components, who is responsible for maintaining the interface in the event of modification of it? Who will take care of the necessary changes to the components on the other side of the interface? (b) If in the interaction between two components A and B, a modification is necessary to correct component A and will take six months, while a modification of component B to circumvent the anomaly in A is possible in two weeks, which is the most "profitable" solution?

– Configuration management must be mastered between all actors (developers and testers regardless of the level of integration), regardless of their company. However, there is no certainty that the tools and methods of configuration management (or version management) are identical between all participants in the design of the system-of-systems. We risk having anomalies that are not reproducible on an environment, only because this environment has a slightly different configuration from the environment where the failure was identified.

2.3.3.1. *Scrum framework*

Figure 2.8 shows the Scrum framework for a project.

The Product Backlog includes all business requests for the product. Often, these requests are in the form of "User Stories". New User Stories can be added (or deleted) to the product backlog according to business needs.

During the Sprint Planning meeting, the Scrum team – which includes all the developers, designers and testers assigned to this sprint – selects the User Stories that will be developed in this sprint. These User Stories will create the Sprint Backlog. User Stories are selected based on the ability of the development team (the Scrum team) to develop them, as well as dependencies on other User Stories and business emergencies or priorities.

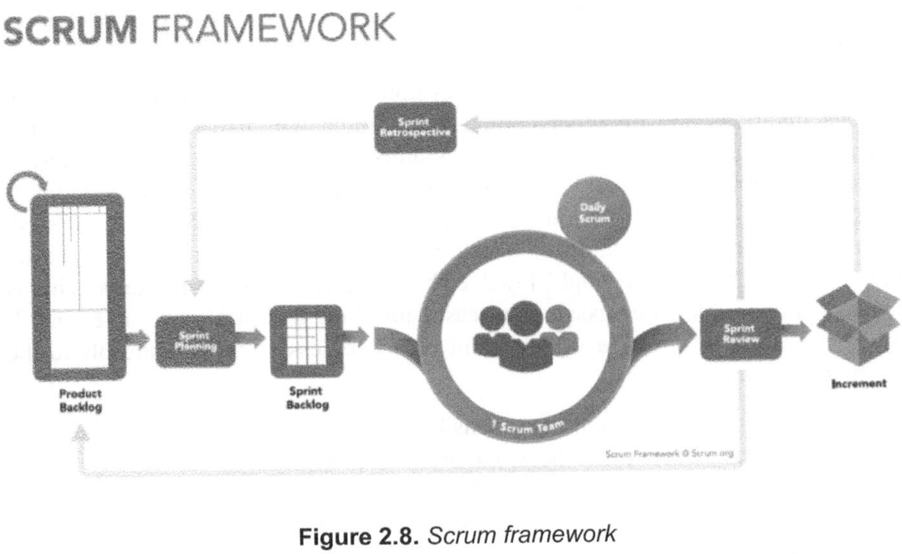

Figure 2.8. *Scrum framework*

During the duration of the sprint – often between two and four weeks – the User Stories will be analysed, developed and tested, until they are considered finished (DONE).

The Sprint Review will ensure that the developments correspond to what the business wanted and the components developed will be part of an increment, potentially deliverable.

Sprint Retrospectives provide feedback on the sprint to improve subsequent iterations.

User stories that do not correspond to business requests, or that require improvement, are reintroduced into the Product Backlog.

Scrum teams generally comprise around nine people, including a business representative (the PO; Product Owner), a Scrum Master (SM) in charge of compliance with the methodology and seven people in charge of development and unit testing.

2.3.3.2. *Nexus framework*

We understand that a team of nine people has only a limited development capacity and that it is necessary – in a system-of-systems – to design an organization comprising several synchronized Scrum teams, each focused on one aspect precise. Two organization schemes deal with these aspects: Nexus (proposed by www.scum.org) and SAFe (supported by Scaled Agile Inc. www.scaledagileframework.com).

Let us see the organization proposed by Scrum.org: Nexus.

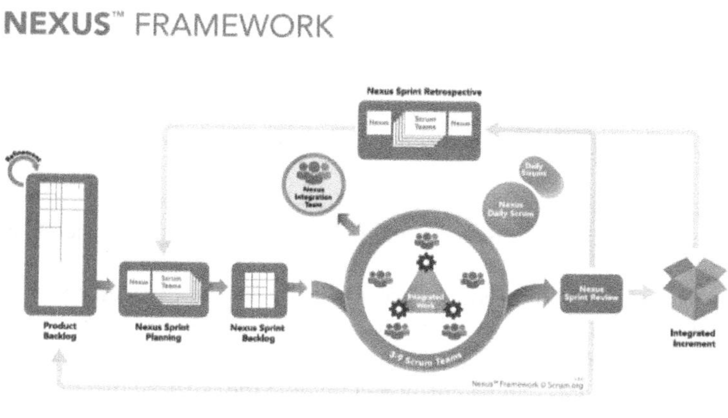

Figure 2.9. *Nexus framework*

This is primarily a scaling up of the Scrum organization, where individual members of a DEV team are replaced at the Nexus level by representatives of those teams.

So, we end up with a Scrum of Scrums, coordinating several Scrum projects in parallel.

Nexus assumes the knowledge of Scrum, and requires the knowledge of several techniques to identify and remove the dependencies created by the increase in complexity linked to the projects to be synchronized. These techniques can be grouped into four areas:

– organize the teams;

– organize work;

– run the Nexus;

– manage the Nexus.

Team organization suggests starting with a small cross-functional team and growing it, then splitting it into several teams, each specializing in a domain or a micro-service, while allowing members to choose the team in which they participate according to their abilities and the needs of the team.

Work is organized by implementing a planning technique that visually maps business objectives and PBI (Product Backlog Items). This allows teams to break down user stories, value them and track them with the goal of reducing or eliminating dependencies between PBIs.

Nexus execution relies on a visual representation of work and its progress, allowing remote teams to visualize the progress of PBIs. Non-synchronized sprint reviews take place to allow non-present stakeholders to provide feedback. The role of the PO can be partially delegated even if the responsibility for the product remains with the PO. Component integration is automated and continuous (CI; Continuous Integration), and acceptance testing is mostly automated.

Nexus management revolves around a tree-like representation of the Product Backlog, Burndown charts of features and performance metrics for each team.

NOTE.– automated end-to-end acceptance tests can be difficult to implement if the test objects are of different natures (e.g. embedded, legacy, web, etc.) and their synchronization requires disparate environments. Managing and synchronizing the various test environments can be costly and involve the creation of a dedicated team, which could become a bottleneck if not properly sized, whether in terms of personnel or resource material.

2.3.3.3. *SAFe Scheme*

SAFe offers itself as a Lean alternative for companies, by implementing several sprint teams working in parallel with periodic synchronization points, a bit like trains that would synchronize in stations. Several variations of SAFe exist, ranging from an essential vision to a vision for large companies. More and more companies are using SAFe as part of systems-of-systems.

To guarantee synchronization of the teams and the projects, the duration of the sprints of each team must always be the same. Any extension of a sprint duration will automatically lead to a desynchronization of the teams throughout the project.

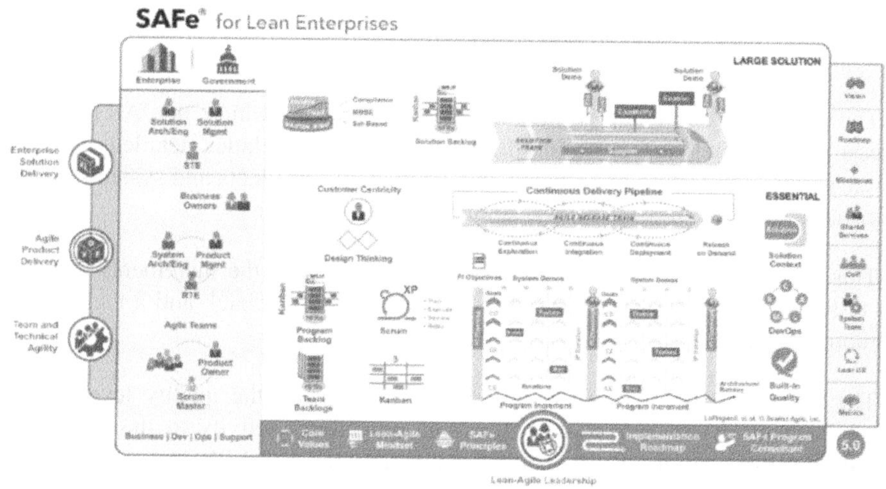

Figure 2.10. *SAFe organization*

The synchronization of the various development teams is carried out by setting up a PI Planning which makes it possible to synchronize the projects between them.

2.3.3.4. *SAFe phases*

2.3.3.4.1. PI planning

PI planning phase identifies and synchronizes the business and technical objectives of the trains. This helps coordinate short-term goals, ensure synchronization between trains, assess teams' ability to meet goals, identify dependencies and the business value each team brings.

It is therefore very important that, in addition to the production teams, the business representatives (the BO, Business Owners and PO, Product Owners), the technical managers (SA, System Architect) and the test team managers (e.g. Test Leads or Test Managers) are present and involved. BOs ensure that requirements meet business needs that add value. The POs ensure the completeness and the level of detail of the requirements, as well as the absence of ambiguities, in order to allow development – then testing – to be facilitated. The test effort – load and deadlines defined by the Test Leads and/or Test Managers – must be included in the execution load and taken into account in the schedules.

The lack of involvement of one or more stakeholders will have negative impacts later.

2.3.3.4.2. PI – program increment

The program increment is the time interval during which the ART (Agile Release Train) plans and performs the increment. This includes detailed analysis, design, coding and testing of software and components, products and systems. SAFe offers an IP duration of 8–12 weeks, comprising four iterations.

This period of fixed duration makes it possible to plan the increments of work and to limit the work in progress, while allowing valid feedback and a measurable addition of value.

The testers are involved in each iteration and ensure the quality level of the deliverables. It is very important to include automation activity in the test tasks. Indeed, the re-execution of tests from previous iterations and PIs makes it possible to guarantee – in a short period of time – the maintenance of the quality of the software, component, product or system.

The annual implementation schedule, for a long-term project such as a system-of-systems, is organized around Program Increments (PI) of fixed duration and planned well in advance. This duration of the PIs is used as a reference for the entire SAFe development project (the entire ART, Agile Release Train) and is split into several increments. The early definition of PIs makes it possible to anticipate what must be delivered at the end of each PI and to provide a common schedule for all project stakeholders. We realize that any change in the duration of an increment will have a domino effect on future increments.

Systems-of-systems often require pre-PI planning preparation meetings and post-PI planning meetings. Pre-PI planning preparation meetings make it possible to coordinate subcontractors – for the data, components, products or systems required as input to the system-of-systems – and to ensure their presence (or their delivery date) to inform participants of PI planning. Post-PI planning meetings allow information from PI planning to be integrated into individual team planning meetings and reach a commonly accepted set of solutions for the system-of-systems in its state before the next PI planning, in the "Demo Solution".

2.3.4. *Lean*

The main objective of agile developments is to provide software that works with the least possible "waste"; many principles of "Lean management" are used and applicable.

As a reminder, the concept of "Lean" and Lean Management evolved from the industry (including Toyota car manufacturing) with the objective to eliminate any waste, whether it is loss of time, loss of materials or any unnecessary action that does not add value to the product. Bug fixes, unused developments, too large solutions, delays due to dependencies, unnecessary processes can all be considered waste. They should all be eliminated.

Applying Lean management to developments means to identify, at the end of each iteration or sprint through a feedback meeting, all activities that have produced waste or delay and to remove them in the future. From this information, continuous improvement activities can be implemented, focusing primarily on the activities that will produce the greatest added value or that will reduce the most loss. After several iterations we obtain optimized processes with the least delays and the least waste.

Lean comprises five principles:

– identify what represents value for the customer;

– describe the stages of the process leading to the creation of the product or service, removing those that do not generate value;

– ensure that the steps follow one another smoothly so as to accelerate execution time for the complete process without generating losses or intermediate stock for software developments (waiting time before a subsequent task);

– allow the client to extract value from each step, rather than preparing actions based on forecasts that may turn out to be incorrect;

– seek continuous improvement achieved only when there is no longer any wastage of resources or value (or time).

To measure the effectiveness of design and production processes, it is important to use metrics and measurements in an iterative and continuous way. Similarly, continuous reporting and constant "self-criticism" will improve the elimination of losses and intermediate stocks.

Lean practices originally come from the manufacturing industry and have been adapted to the IT industry. Several Agile practices are linked to the Lean approach, such as KANBAN and aspects related to the optimization of production flows (e.g. just in time).

NOTE.– The principle of "intermediate stocks" comes from the industry where the execution of an operation on a machine (a batch) is only profitable if the actions of preparation and finalization can be divided by the number of parts processed by the machine once. This leads to the production of intermediate stocks before the machine (e.g. to prepare an economical batch) as well as after the machine (e.g. such as to allow cooling of the parts after treatment). The smaller the batches, the higher the preparation and finalization costs with regard to each part. In the context of IT development, this implies that it is not necessary to wait until all the requirements are defined before studying them in a single batch, but that it is possible to start processing the requirements as soon as they are elicited.

Among the "losses" to avoid identified by Lean in Poppendieck and Poppendieck (2015), we have everything that does not provide added value such as:

– partially completed work;

– unnecessary processes or paperwork;

– unused additional features;

– moving from one task to another by assigning the same person to several tasks;

– waiting and delays in deliveries;

– the time required to obtain information;

– defects;

– management and management activities.

For Lean to be expressed successfully, the mentality of participants and organizations must be "positive" and anomalies or failures must be seen as opportunities for improvement rather than as punishments. With a positive mentality, each anomaly will be investigated impartially and will allow continuous improvement. On the other hand, if the anomalies – and the shortcomings that result from them – are seen as things to be avoided and repressed, they will often be denied or hidden and neither the participants nor the organization will learn from these shortcomings. See also Syed (2015).

The objective of Lean is to eliminate waste and limit the loss of value as much as possible. So early detection (to avoid increased costs) and root cause analysis (to prevent these bad practices from being repeated) will be necessary.

2.3.5. *DevOps and continuous delivery*

DevOps is a Lean extension of the Agile implementation whose objective is to reduce the friction between DEVelopment and OPERATIONS (Sacquet and Rochefolle 2016) while mastering "the construction, the evolutionary maintenance and the large-scale operation computer systems that must be able to be modified quickly and without failing". As this is part of a Lean perspective whose objective is to improve the performance of an organization by controlling productivity, quality, deadlines and costs, it is obvious that these activities must take into account the tests (quality) at all levels, whether it is quality of deliverables or quality of processes, even automation of processes.

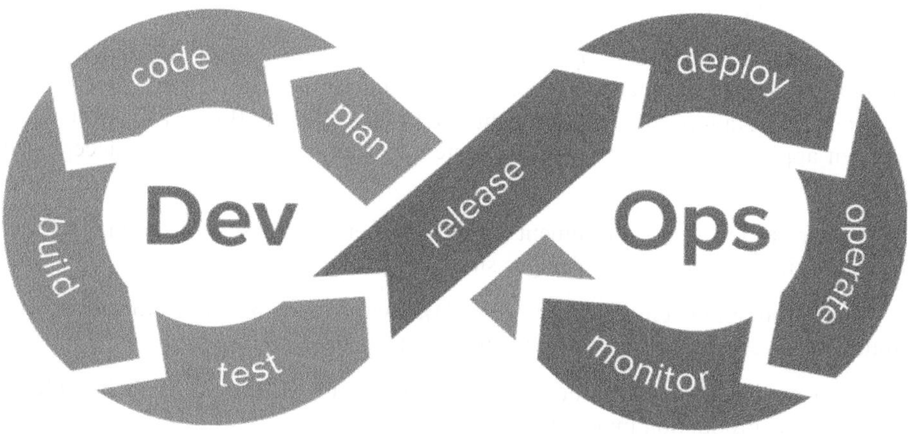

Figure 2.11. *DevOps representation*

Among the Lean elements of DevOps, we have:

– the value chain, where productivity must go until the production of the designed product, in order to bring added value to the company;

– the need to ensure the absence of regressions very quickly, which implies implementing intensive test automation, from unit testing (e.g. via TDD or static analysis) to testing of acceptance and verification of the effectiveness and relevance of the tests (see also volume 2, section 10.6.3).

Various elements must be considered when considering a DevOps solution:

– Tests must be automated, from unit tests to end-to-end tests, including integration tests and system tests. This requires representative test environments available to development and test teams, automatically updated so each build can be tested. Depending on the number of test levels, and of the workload to keep the automated tests up to date, the executions of functional tests and regressions should be frequent (e.g. at each build or daily) and easily maintainable.

– Deployment activities, and rollback activities in the event of a problem, must also be considered and validated to avoid problems with putting into production or moving from one environment to another.

– If a deployment without service interruption is necessary, it will be necessary to provide multiple environments with load balancers, allowing a certain redundancy and, for example:

- a development environment (DEV), optionally several if parallel developments take place,

- a test environment by level with the various systems, applications, components or products necessary for the execution of the tests. Often, we will have a system integration test environment (SIT) to validate flows and exchanges between applications and at least one system test environment or user acceptance test (UAT),

- a pre-production environment (PPROD) where systems validated in UAT can be tested with volumes and exchanges similar to production,

- a production environment (PROD), with load balancer to allow the production or correction of software components at any time,

- an environment reproducing the production (VALID), so as to be able to reproduce the production incidents which could occur.

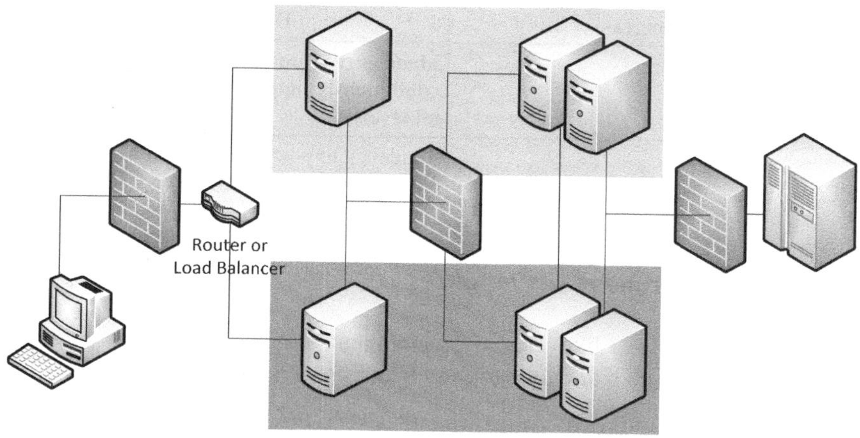

Figure 2.12. *Redundancy and production platforms*

This multiplication of environments will have an impact on configuration management – including test data – as well as on tests, the processes to be implemented in the event of fault detection (corrections and tests on several versions) and the needs to revise all the stages of realization/test to improve them. Increasing environments will also increase costs.

Continuous delivery is the basis of the Agile and Lean development philosophy, as defined in DevOps-type environments, where the application is periodically and regularly updated and delivered in production to users.

Reminder: in Agile developments, the delivery at the end of the iteration of an application that can be used by users is planned. For a development over several iterations, it is obvious that the intermediate deliveries are incomplete and only the last iteration includes all the modules.

Considerations for continuous delivery include:

– distribution activities, deployment, transfer of information from an old version to the new version (not to mention backtracking);

– training and coaching activities for end users on the new features of the application. If every new feature requires training, in the case of releases at the end of each iteration (e.g. three weeks), users will be forced to participate in training every three weeks.

Advantages	Disadvantages
Frequent patching to users	Mandatory regression tests, almost continuously, implying automation of tests at all levels
	Tendency to postpone corrections to subsequent iterations

2.3.6. *Agile development challenges*

Agile developments, like other development models, are also subject to challenges, some of which are not currently being properly addressed:

– Limited understanding of Agile principles seen through the prism of relatively short-term development. The Agile Manifesto and the Agile principles are not binary, and mention preferences; we can have an answer range in the application of agility, depending on the business domains and the constraints of the organizations.

– Continuous delivery is an interesting principle, including several challenges, such as (1) when the delivery involves a need for user training or (2) the need for regression testing which increases with each new sprint. In the first case, a continuous delivery will lead to a need for continuous training (heavy and problematic) which must be anticipated for the testers and for the users. In (2), the load required to manually and continuously check the impact of changes on the operation of the system will increase linearly with each sprint and will become too great in relation to the expected benefits if the tests are not automated. During continuous deliveries, or frequent deliveries, it is imperative to implement – simultaneously – automated tests.

– Automated tests and manual tests: test automation is often linked to Agile methods. Automated testing is required with TDD, BDD and ATDD techniques. However, these techniques are not really tests, and are only useful for finding regressions. Manual tests alone allow us to verify and validate new functionalities, human testers being able to find defects unlike automated tests. As mentioned before, as the sprints progress, there will be more and more features to test, therefore a greater test workload, although the principle is based on constant duration iterations. We will therefore be required to implement transverse automated tests, making it possible to verify the absence of regressions in the end-to-end functioning of the system-of-systems. The solution of automating the tests can slow down the impact but cannot cancel it in the long term. Another solution would be to gradually reduce the scope of the tests to be performed, to keep a constant workload and duration, but this will increase the risk of regressions given the deletion of tests.

Advantages	Disadvantages
Allows us, if the implementation is correct, to provide a constant evolution of the functionalities	Requires a fundamental paradigm shift for customers and delivery teams (combined dev and test)
Meets business needs in terms of managing evolving needs	Requires complete documentation to compensate for the turnover of production teams and to comply with legal obligations
The use of Gherkin-type languages and Pipelines facilitates test generation and to run them automatically during builds	Designing and automating tests for new User Stories typically cannot be done during the User Story Design Sprint
	The increase in the number of tests – automated or not – will lead to increasingly heavy test load
	Initial testing strategy must consider manual tests (or tests to be run only periodically) and tests to be run at each build

2.4. Acquisition

In the previous sections, we have looked at the design aspects of a system from scratch, but in a system-of-systems, we will reuse existing components within the system-of-systems architecture. This also involves acquiring components already developed and ensuring that these components meet the needs of the system-of-systems, just as we will be obliged to deal with obsolescence of components or systems.

This involves defining, at the level of the system-of-systems and each of its components, objectives in terms of quality attributes such as those defined in Chapter 5. One way to do this is to identify operating paths (nominal and exceptions) of the system-of-systems, defined based on the architecture of the system-of-systems, and to mention measurable objectives – among others for reliability, availability, capacity, bandwidth and response time. To define these objectives – and ultimately the SLAs of each of the components – we will have to carry out a detailed analysis of each of the stages of the functional paths of the system-of-systems, to identify the components or nodes involved and to specify the SLAs of each of these nodes of the system-of-systems.

The acquisition of system-of-systems subsystems and equipment must also consider the test environments that will qualify the system-of-systems, whether during the initial design or during future developments. The qualification, verification and validation environments will require the installation, use and configuration of acquired subsystems and equipment. These acquired components will have to be managed in configuration like the developed components mentioned in the previous sections.

2.5. Maintenance

Software must evolve, whether to correct defects, provide upgrades or meet regulatory obligations. All these evolutions may generate defects and therefore must be tested. Maintenance tests, planned or urgent, are divided between customer and supplier sides. Various questions arise during maintenance, which are also applicable for the correction aspects of systems-of-systems:

– Should they be corrected immediately or grouped with evolutions distributed less frequently?

Fixing immediately is necessary for critical fixes (e.g. security), but involves a disruption of normal delivery processes and frequently a reduction in testing activities related to these fixes. The integration of corrective maintenance with evolutionary maintenance makes it possible to test both types of maintenance more systematically.

NOTE.– This mode of distribution may vary depending on the type of organization chosen by the company (e.g. DevOps).

– When modifying certain systems or system components, areas for improvement can be identified. Should they be implemented directly, even if it is not planned?

Some development modes, such as Agile methods, offer "refactoring" iterations whose purpose is to improve the code without adding new functionalities. More traditional development modes can also identify potential improvements in subsystems or software. Should these improvements be implemented or not? At first glance, such an implementation is reasonable and cost-effective. However, any refactoring, any modification of the code bringing new functionalities or not, must be justified because it will require testing efforts to ensure the absence of regressions.

– Should we directly correct the system in question, the one that contains the defect, or should we ensure that the defect is controlled only in the user interface, to reduce test and certification times?

Let us take an example: in aircraft, a guidance system exists which is based on the detection of radio beacons, which emit information from specific locations, such as lighthouses for maritime navigation. By identifying the position of the aircraft relative to several beacons, the pilot can determine its position on a map. This navigation system depends on a detector which indicates – for each radio beacon – the direction (in degrees) from the aircraft. If the detector indicates a 180 degree opposite position from the actual direction of the beacon, the obvious solution is that it is this equipment that needs to be corrected. However, if the deviation (here 180°) is known and constant, it may be possible to remedy this failure by correcting the value during the display. In absolute terms, a fault is introduced into the display software, linked to the fault in the beacon detection system, but the final display is correct. If the correction and qualification period for the display is one month and the correction and qualification period for the beacon detection system is six months, it may be of interest – for the aircraft designer – to change the display first and wait for the beacon detection system to be corrected and requalified.

– Should we take advantage of maintenance to reduce technical debt in existing applications?

This solution gradually increases the quality of existing applications by using maintenance activities – evolutionary or corrective – to improve the functions, classes, subroutines and other components undergoing maintenance. The components are thus improved one after the other according to the changes requested. This will result in improving the components of the application bit by bit, without major impact in terms of time or budget.

– Once the software is in maintenance, how to integrate it into the higher levels of the system-of-systems? Are updates pushed to users (PUSH) or requested by users (PULL), and how will integration be checked at each level (impact on interfaces and other systems)?

As mentioned earlier, it is necessary to ensure the quality of any changes or updates to the system. This involves regression testing. In a PULL solution, users can decide whether to implement an update, planning the test phases according to their schedules. In a PUSH solution, users are forced to schedule these testing phases according to the vendor's schedule; this schedule may not be in line with yours.

2.6. OK, what about reality?

Software applications and systems-of-systems can be thought of as trees, which evolve little by little by adding new features, or we can think of them as bridges where, until the bridge is fully completed, it is impossible to use. In the first case, Agile development modes are very suitable, while in the second case, sequential development modes are more suitable. Next come trends, fashion and maintenance.

Fashion is moving towards agility, with the generalization of Scrum and SAFe. In large companies, using CMMI, ITIL or PRINCE2, the trend is still often towards sequential development, but agility is developing to meet the needs of companies for increased responsiveness to the market. Whatever the initial development cycle, once the system goes into maintenance, it is possible to switch to a DevOps mode with the implementation of Lean recommendations.

Do these various cycles of development exist in reality? According to the companies, the changes are made little by little (pilot project followed by a sequential generalization after validation of the pilot) or by good techniques or technologies implemented according to an Agile philosophy. As a result, legacy-type systems coexist with more modern development cycles, which leads to parallel evolutions, where some systems are iterative, others agile and others purely sequential.

These developments often represent a different vision between the business, on the one hand, and the objectives of the development teams, on the other. ERP- or CRM-type systems such as SAP (in production) or SAB (in banking), or even others such as SalesForce, JD Edwards, etc., involve a specific mode of organization, often sequential. Many organizations use sequential development cycles for their legacy systems or for planning their IT strategy, with milestones, annual budget decisions. The transition from a traditional mode to an Agile mode should not be done on a whim, nor without carrying out a detailed reflection on the impacts linked to a change in the development paradigm. Among the elements to be taken into account, we should cite the positioning of software testing and the ways of validating the quality of software development, taking into account the large – and growing – volume of testing activities (increased effort to automate, increase in scope and complexity of systems and their interactions).

What we often see is a mixture of systems having to interact with each other, but according to different development systems:

– Legacy, ERP and/or CRM systems, as well as all systems requiring significant design (or deployment) time, often follow a sequential or hybrid development mode (higher levels are sequential, lower levels are pseudo-Agile).

– Integration of new systems (ERP, CRM, Business Intelligence, Artificial Intelligence, Data Lake, etc.) seems, for some companies, to be a cheap solution meeting the needs of rapid development of their market, and a guaranteed source of profitability future. The publishers of these solutions boast of the benefits of their solutions, but forget to mention the costs and durations of setting up these solutions, or the impacts that these solutions could have on the legacy applications in place.

– Some systems are pseudo-Agile, in that they start with all the specifications and end with a user acceptance phase, and the development phases are sequential or v-shaped or Agile, but the releases are not provided to users at the end of each iteration.

– Smaller systems are developed in an almost Agile way (using Scrum, KANBAN, XP, etc.) but the interfaces of these systems with legacy systems (CRM or ERP), as well as the Legacy developments associated with these systems, are carried out sequentially, with clear phasing allowing controlled production launches.

Such organizations are difficult to master and optimize. Testing, review and continuous improvement activities are only implemented gradually, depending on the systems and minimizing the impact on other systems.

As mentioned by Ford et al. (2021), another element to take into account is the evolution of system architectures, with the current trend towards decoupled solutions based on services and responding to events (event-driven architecture), including microservices and coupling via APIs, as well as the development of massively parallel architectures, even the emergence of architectures based on quantum systems. These new architectures – as well as the synchronous or asynchronous exchanges they imply – will impact the verification and validation activities of complex systems and systems-of-systems.

New development trends are emerging like MBSSE for model-based systems and Software Engineering based on ISO/IEEE (2021). Their reference model is structured around four process groups (Plan MBSSE, Build Models, Run MBSSE and Support Models). In the context of systems-of-systems, the concepts of RAMS (Reliability, Availability, Maintainability and Safety) and RAMT (Reliability, Availability, Maintainability and Testability) are concepts to be considered, account from the design of the system-of-systems and each of the systems. The systems developed based on MBSSE will require both verification and validation of the models generated, as well as compliance with the recommended processes, verification and validation of the development and engineering actions carried out and the products delivered. In such contexts, it seems that traditional development cycles are better suited than Agile cycles.

In this chapter, we have mentioned the terms "added value" and "customer" several times. It is necessary to recall the definitions of these two terms because their interpretation influences understanding: a customer is a person – legal or physical – who brings new money to the company. Therefore, any exchange giving rise to invoicing between the services of the same company cannot be considered as a "customer–supplier" relationship insofar as it does not bring fresh money to the organization. This vision of the "customer" makes it possible to define the elements bringing an increase in value for the customer and to dissociate them from those bringing value to the organization. For example, an improvement in the user experience (UX for User eXperience) will make it easier for the customer to use, while an improvement in maintainability will bring a potential reduction in development and code correction times, but only in the future. This reduction in maintenance effort will bring "added value" to the organization in the sense that it will reduce expenses, which is one of the objectives of Lean.

3

Test Policy and Test Strategy

Within a company, software testing only exists within the framework of implementation projects. It is therefore necessary to define their positioning in relation to the company's processes, objectives and strategies to implement them. This revolves around two main documents:

– the test policy, transverse to the whole organization;

– the test strategy that will describe and detail how the test policy will be implemented for a given project.

It should be noted that within a group made up of several companies each having a separate information system, it is rare for the policy and/or the test strategy to be identical for all. We will then have specific policies for each company.

Test policy and test strategies can be differentiated according to the type of system (new or under maintenance). The testing policy defines why we test in the organization and can vary according to the types of systems to be considered, while the test strategy will define how the objectives defined in the test policy will be achieved.

3.1. Test policy

Test policy is a high-level document defining the benefits that the company wants to derive from testing, as well as the application scope, the main processes and the metrics to measure the achievement of these objectives. A company's test policy defines, for the whole company, "why" tests are performed. There are several excellent reasons for testing: to find defects, to improve the user experience or the development process, to ensure quality throughout the design process, to reduce design times, etc. Depending on the objective(s), test activities will be organized differently and the reporting will need to be adapted.

As the development of systems-of-systems impacts several organizations, it will be necessary to ensure that the level of quality of the constituent parts of the system-of-systems is of adequate quality. Each development of a system must implement the processes and actions necessary to achieve the objectives defined in the test policy. A test policy is therefore important in that it makes it possible to associate test activities – as well as continuous quality improvement – with each organization's activities and thus legitimize the recommendations for improvements identified during tests and through analysis of the faults found.

3.1.1. *Writing test policy*

As the test policy must apply to the whole company, it should be signed by a person of sufficient hierarchical level (e.g. CEO, general management) so that the guidelines defined are applicable to all departments of organization as well as co-contractors and sub-contractors of systems-of-systems. In order to be applicable and enforceable by subcontractors, it is important that this testing policy be referenced as "applicable" in the contractual documents, and that it be provided as an appendix to the contracts. Also, as this testing policy should apply to all departments, it is recommended to circulate drafts and get feedback from design teams before finalizing and publishing this policy of testing.

The test policy will often be drafted by a test manager or a quality manager, on behalf of the line manager who signed this test policy.

3.1.2. *Scope of the test policy*

By default, the test policy shall apply to all development projects in the organization. Some projects may justify not implementing all the recommendations present in the test policy (e.g. very small projects or R&D projects not intended to be maintained over a long period, or even "legacy"-type projects for which a policy based on new developments has no interest, etc.). In such cases, a partial – and incremental – implementation of the test policy may be necessary to avoid a threshold effect, pushing project managers to remain below the trigger threshold.

When the organization is made up of several geographically distributed establishments – within the same country or several – it may be necessary to ensure that the testing policy is consistent and understood in the same way regardless of the location. This can be difficult depending on the specialization of each geographical entity, and the fact that a test policy will first be implemented in one location before being generalized throughout the organization.

It is also important that the test policy takes into account – and is accepted by – subcontractors and external personnel working for the company.

3.1.3. *Applicability of the test policy*

It is important to note the scope of the test policy: it is unrealistic to apply a very heavy test policy on R&D projects or on projects below a certain workload. To avoid that, the test policy is considered only as a collection of good intentions; it is necessary to define a reporting process with periodic metrics and measures to ensure correct implementation of the test policy, and to measure continuous improvement.

3.2. Test strategy

The test strategy implements and details the test policy on a system or on a system-of-systems. This involves the coordination of actions, activities and test methods to ensure that the system or system-of-systems meets the objectives defined in the test policy. Thus, we shall first define these objectives and how to measure them in terms of benefits for the organization. The objectives can be, for example, the delivery of a system-of-systems having no defects ("zero-defect") or the supply of a system-of-systems at a defined date with a given number of functionalities and a defined level of quality. Another aspect to consider is the extent of the scope of application of the strategy: does it apply to new developments, maintenance – which will require more regression testing – or does it apply to one or more components, which may lead to the implementation of transverse and/or end-to-end tests in the system-of-systems.

The test strategy for a system-of-systems must show how the company's objectives for the operation of each of the individual components, their interactions and the complete system-of-systems will be verified and validated, including information on how compliance with requirements will be demonstrated. This involves identifying the places in the system or system-of-systems where defects could be introduced, and the phases during which these defects must be identified and fixed. A test strategy must also show the differences between standard management processes and processes applicable on the project, and how they fit together (defect management, release management, configuration management, reporting, etc.).

As the test strategy should explain the logistical aspects of system-of-systems testing, it will be necessary to address aspects of test environments, transfer milestones for a software application (or a product, a system, etc.) from one environment to another (including the prerequisites and the definition of an adequate level of completeness of the tests carried out) and communication between the actors.

As mentioned before, the test strategy references and implements the test policy, but it must also consider the development models (V-cycle, Agile, Scrum, SAFe, etc.), systems or systems-of-systems to be produced, commercial targets (which could impact the need to translate the user interfaces), as well as the reality of the company's business. As the purpose of testing is to find defects as early as possible, it is important to identify the phases of introduction of defects according to their typology and which test activities will be able to identify these defects as early as possible. One aspect to keep in mind is that testing must be simultaneously effective (find defects) and efficient (find the defects without waste of effort). Some checks (e.g. translations) may require human intervention and therefore manual testing, while others can be automated (e.g. regression testing).

A possible test strategy would consist of identifying the stages of development and associate with each stage one or more test or review activity, each focused on the type of defects that can be detected at this stage. Milestones for deciding whether to continue or move on to next steps (sometimes called Go/NoGo meetings) can be located between steps.

If the development cycle is sequential, the strategy can be easily described in a work breakdown structure (WBS) where each action is clearly identified, with its prerequisites, deliverables, etc., or by a fishbone diagram (Ishigawa diagram).

If the development cycle is Agile, it will be necessary to correctly define minimum exit criteria from the grooming and sprint planning phases, as well as to ensure that the definition of "Done" is correctly understood and applied by all. In the sequence of tasks (e.g. KANBAN board of progress of a user story), focused test actions – or test phases – are indicated (e.g. carry out a code review before saving its coded). A possible solution would be to design the user story tests simultaneously with their design, but to automate them on the following sprint (if automation is necessary).

In the majority of cases, it will be necessary, for each activity, to define and identify the prerequisites, whether it concerns the presence of features, user stories or requirements, environments, data, settings, equipment, flows from and to other components or applications, etc.

This way of organizing the focus of the tests on the typologies of defects introduced in the previous phase makes it possible to design successive filters, each responsible for finding specific defects. If these filters are not put in place, the faults can only be identified during the subsequent test phases, that is, at a time when the correction (and retest) times will be shorter.

3.2.1. *Content of a test strategy*

A test strategy shall explain:

– How the system or system-of-systems will be tested, for example, via a breakdown into domains or projects dedicated to specific aspects (or quality characteristics), or by a breakdown into levels, or a breakdown into components, products or systems (individual tests of the elements of the system-of-systems and then integration of these elements with each other) and by testing their relationships and the messages exchanged.

– The organization of system-of-systems tests in the context of the evolution of versions of the organization's systems, its constraints and dependencies.

– The levels of integration and verification, and prioritization criteria.

– The teams, departments or parts of the organization that will contribute, decide and/or be informed of the evolution of the system-of-systems tests.

– Which quality characteristics are most important for the system-of-systems.

– How each of the test levels will improve the quality of the component, product, system or system-of-systems and provide useful information.

– How the components will be managed in configuration, including the management of developments, corrections of anomalies, decision-making committees (CCB for Change Control Board).

– The organization (who does what), reporting and governance bodies of these system-of-systems tests.

A test strategy must:

– reference the objectives defined in the test policy and applicable to the system-of-systems on which this test strategy will apply. If necessary, reference the test policy with its version;

– identify or recall the goals of the system-of-systems for each type of end user;

– define objective criteria (metrics) to measure the level of achievement of the stated goals of the system-of-systems and of each of the systems or subsystems that compose it;

– identify the risks to anticipate and mitigate, on the project and on the product;

– identify the methods of proof to be put in place to verify and validate the requirements on the project, according to contractual and regulatory needs;

– identify the environments to use and the criteria for moving from one to the next;

– identify the steps necessary for the implementation in order to achieve the set goals. This means:

- the prerequisites, entry conditions and exit criteria associated with the test activities,

- delivery activities (configuration management),

- defect management processes,

- implementations (publication, production, etc.) and distributions on the must also be taken into account;

– define the test methods to be implemented, for example, according to the level of risk (see section 4.1).

A test strategy detailing and implementing the test policy aspects of each company shall include:

– the testing techniques to be used (e.g. depending on the level of risk);

– the scope and level of independence of the tests;

– test environments (including version evolution processes);

– the objectives in terms of test automation;

– how to confirm defect fixes and ensure the absence of regressions;

– measurement of test activities (metrics, frequency, reporting);

– control of test activities based on measurements and reports;

– defect management;

– delivery management (e.g. in the case of continuous deliveries);

– test reusability and configuration management;

– applicable standards;

– procedures for integration, reception at the beginning of an activity and transfer to a following activity.

Each strategy involves assumptions or hypotheses that will influence the outcome. It is recommended to list hypothesis and assumptions.

3.2.2. *Test strategies and Taylorism*

As you may recall, Taylorism consists of the separation of complex tasks into simpler optimized individual tasks, the objective being to allow simplification and optimization of these tasks. As part of a test strategy or an approach to test activities within an iteration, this amounts to splitting test activities, sequentially or in parallel, to optimize effectiveness and efficiency of these testing activities. We could thus define the activities to be carried out at each test level, according to the design phase of the system, the actors, etc. This splitting into smaller tasks is the most frequent mode of operation and is correct for the validation of systems during their integration tests. This translates to a WBS or TBS for Test Breakdown Structure.

Within the framework of a sequential design, a division of the test tasks according to the system proposed by Frederick Winslow Taylor is justified because the system is hierarchically composed of smaller components. This division allows, among other things, an assignment of verification activities (e.g. reviews, static analyses) and validation (component, integration, system, system integration and acceptance tests) to each of the levels of test and also after each achievement task. So, the specialization of testers and the automation of certain test tasks can be considered, but it will be necessary to ensure – and measure – that each of the test activities has been correctly carried out, since each activity is focused on the detection of a type of defect.

In Agile design, as well as in the case of corrective or evolutionary maintenance of existing systems, the test activities relating to part of the code will have to cover almost simultaneously all the verification objectives (reviews, static analysis, etc.) and validation for all test levels. The time required to analyse the impacts (side effects that can generate regressions), design, implement and run the tests, then manage the failures (identification, correction and confirmation of these) is often not sufficient for these actions are carried out in a single sprint – a single iteration – by specialized testers each in their field.

The impact on the test strategy is significant and will affect the level of knowledge of testers and developers, the need for simultaneous availability of test environments usable for the tests of each level, the methods of transferring the system under test from one environment to the next and the measures to validate the completeness of the completion criteria of the achievements (both the aspect of DOD and the measurement of the velocity of achievement). For example, we will no longer be able to have testers specialized in one area or another, identifying defects sequentially, but we will need personnel with multiple knowledge (both in code design methods, in networking, in security, performance, etc.) able to simultaneously process the various levels and types of test.

3.2.3. *Types of test strategies*

Four main test strategies are possible for an application. Some are better suited to systems-of-systems than others:

– analytical strategies;

– methodical strategies;

– reactive strategies;

– consultative strategies.

These strategies can be combined and adapted to take into account the constraints and particularities of the organization and the system-of-systems to be tested.

3.2.3.1. *Analytical strategies*

We speak of an analytical strategy when the product, the project and the documents that describe them are analysed, and the result of this analysis determines the tests to be performed. Analysis activities take place before test design and implementation activities, and before test execution.

Analytical test strategies are well suited to sequential development cycles, and implement test methods based on requirements and risk analysis.

Analytical strategies have many benefits:

– It is possible to map the test needs to the requirements, to the risks, to the components described in the analysis documents. The repository (the test base) is generally well documented, and it is possible to create a detailed traceability from the requirements to the tests.

– Identification of elements to be tested allows easy measurement of test progress. Traceability between the requirements and the tests enable the calculation of the percentage of coverage of the requirements by the tests. This provides clear information on what works, what doesn't, what has been tested and what still needs to be tested.

– Analysis of the test basis makes it possible to anticipate defects and avoid their introduction. Review of requirements and other design documents (architecture, specifications, etc.) enable the identification of defects or risks and allows the definition of actions to eliminate or limit these early in the project.

Analytical strategies have drawbacks (similar to sequential life cycle):

– Analysis, mapping and prioritization of requirements, assignment of requirements to components, architecture and other initial activities may lead to scope bloat, or at least to trying to add too many functionalities to the system. The result may become impossible to design and develop within budget.

– Reviews and inspection of requirements and architecture may take a long time and need multiple iterations before signing off. This increases the duration of the project.

To be successful, analytical test strategies must consider the following factors:

– For document-based strategies (e.g. requirements, specifications, design, contract, etc.), test teams should receive these documents prior to the test analysis activity. Documents may be drafts, and need to undergo revisions, but should be of sufficient quality to uncover test environment requirements and needs.

– For strategies based on the analysis of stakeholder needs (e.g. risk analysis), the test team must obtain the participation – in the analysis and risk assessment activities – of all the stakeholders including business and technical stakeholders. This should identify the test conditions and environments as well as the level of risk associated with each test condition.

Analytical strategies are sensitive to changes in requirement documents and the test repository. Any changes will have an impact on testing activities. This impact will grow as the activities progress. Some sequential development cycles limit the possibility of changing requirements during the project (e.g. spiral cycle: section 2.1.3).

Agile development cycles assume that requirements are stable during the sprint – the iteration – in which they are completed. Requirement changes are only considered from the following sprints or iterations.

3.2.3.2. *Methodical strategies*

In a methodical testing strategy, test teams follow a defined set of test conditions. The origin of these conditions can vary, and could come from quality standards (e.g. ISO25010) or from test checklists built up over time, or from lists compiled based on their experiences, coming from the business domain or characteristics of the component itself.

Methodical testing strategies use test conditions that do not vary across iterations or projects, although updates to these conditions may occur.

The advantage of a methodical strategy is test stability: once the test checklist for an attribute is defined, it does not have to change and will be reused regardless of the project or product. Checklists are typically defined at a logical level, which allows their completeness to be assessed.

The disadvantage of a methodical strategy is that the selected test conditions may not reflect the specific project's needs.

To succeed with a methodical strategy, the following elements must be present:

– Test conditions, checklists and use cases must reflect what the system and the users actually need. They must be complete, correct and accurate, therefore periodic updating of these test conditions must take place.

– As updates of the test conditions, checklists or use case will take place with a relatively long periodicity compared to an analytical strategy, this type of strategy is effective when the component to be tested is stable and varies only very little over time.

The main risks associated with a methodical strategy are related to the obsolescence of test conditions, checklists or use cases, if the component to be tested must evolve. In this case, it is necessary to adapt the obsolete elements and replace them following the analysis of the components and the risks that may be associated with them.

3.2.3.3. *Reactive strategies*

A test team follows a reactive strategy when it reacts to the system or software delivered to the test team. Reactive strategies can be used regardless of the development cycle and are characterized by:

– test design and implementation primarily in parallel with test execution;

– a test plan and a mainly dynamic test approach, evolving according to changes in the product to be tested;

– the use of techniques mainly based on the exploration of the application and the defects identified previously, with the help of heuristics, attacks and defect taxonomies;

– light and informal documentation and processes.

The benefits of reactive testing strategies are:

– identifying defects that other strategies might have overlooked;

– increased defect detection effectiveness resulting from reduced documentation effort and rapid test evolution;

– continuous adaptation of testing efforts. Information gained during testing influences subsequent testing and testing teams can focus on the most productive testing areas;

– a rapid and robust response to any changing context and situations, as test conditions can be defined even when the test basis and design documents may be missing or incomplete.

The following risks and disadvantages (non-exhaustive list) must be considered to allow reactive testing strategies to be carried out successfully:

– if the skills or experiences of the testers are not applicable for the product to be tested, the tests will be superficial and will ignore – perhaps important – areas of the application. Important faults could thus be ignored;

– impossibility of anticipating faults and therefore of avoiding their introduction into the system;

– the absence of documentation – the default solution of such a strategy – can lead to a level of coverage of requirements that is unknown. The use of test charters (see exploratory tests) makes it possible to limit this.

3.2.3.4. *Consultative strategies*

A consultative strategy is one where the test team asks a small number of key stakeholders – the consulted stakeholders – to select the test conditions to cover. The testing team will then define the specific tests to cover these conditions. The main difference between this strategy and the analytical strategies is that the test teams only cover the test conditions defined by the consulted stakeholders.

When a consultative strategy is properly implemented by the test teams, it has the benefit of providing stakeholders with the appropriate level of testing effort on the items those stakeholders have defined. Proper execution involves test teams designing, implementing and executing the correct tests, based on specified test conditions.

A consultative strategy only brings benefits when stakeholders define the right areas to test, whether technical or business. The relevance and level of detail of the test conditions directly affect the success of this strategy. If the objectives of the test conditions are not correct, the test teams will focus on less or unimportant areas, leaving out critical elements.

To be successful, a consultative strategy must:

– include consultation of two or more stakeholders, even if they have different opinions on what should be tested, or the level of testing to be implemented. This implies that the testing team must identify and resolve these differences of opinion (e.g. involving another stakeholder, or a candid);

– ensure that all stakeholders are consulted, so as to prevent a stakeholder from feeling aggrieved – because not consulted – and reacting negatively towards the test team;

– be used in combination with other testing strategies and include coverage metrics that will help ensure the testing effort is scaled correctly.

3.2.4. *Test strategy and environments*

The test strategy must consider the test environments associated with the development models and test levels implemented. Let us take an example.

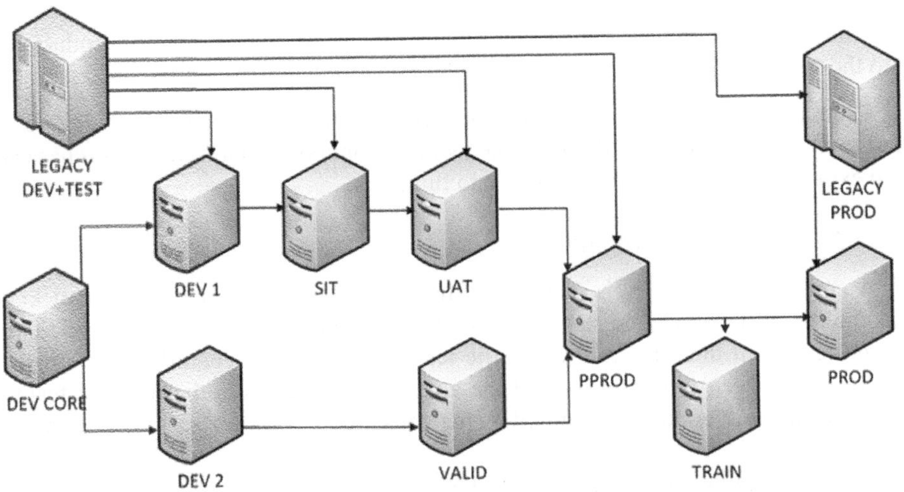

Figure 3.1. *Development and test environments*

If we consider environments like those presented in Figure 3.1, we see the complexity of the exchanges between the environments. We have the lower branch (DEV CORE → DEV2 → VALID → PPROD →PROD) which focuses on urgent fixes, while the upper branch (DEV CORE → DEV 1 → SIT → UAT → PPROD → PROD) focuses on developments of new features. The LEGACY DEV+TEST environments here follow a development and validation that integrates with DEV 1,

SIT and UAT activities, with a link to the PPROD to check the volumetric and response time aspects, then a deployment to LEGACY PROD in production.

Each environment should include software, components, products and systems representative of the level of testing covered by the environment (integration, validation, acceptance, etc.). These environments should all be synchronized, but such synchronization is difficult to achieve.

3.3. Selecting a test strategy

Each of the proposed strategies provides a solution adapted to an environment, to a context.

A one-size-fits-all solution will not meet your needs. You can combine multiple test strategies and multiple test methodologies to design the optimal test strategy for your project. The strategies and methodologies to be combined may vary according to the components of the system-of-systems and according to the stages of development. A good strategy must be able to identify the most frequent sources of defects in order to propose methods for identifying and eliminating these defects.

How to select the one that will meet your system-of-systems needs?

3.3.1. *"Completeness" of the strategy*

A test strategy must be articulated to consider the overall end-to-end organization for the system-of-systems quality. The goal is to be effective and efficient, and it is more profitable to avoid generating defects rather than trying to filter them out. This involves the analysis of development methods and phases as well as the implementation of metrics and reporting at each stage.

In order to optimize the early identification of defects, it is necessary to verify and validate the elements at the origin of the design of the software code and the systems of the system-of-systems. This involves the implementation of static (reviews, inspections, code analysis, etc.) or dynamic (execution of test classes in TDD, execution of unit and integration tests) test activities for each of the systems in the system-of-systems.

One way to ensure the completeness of the strategy is to ensure that each test level or each test phase covers test objectives complementary to the other test objectives of the other levels and phases, avoiding testing twice the same thing.

Because system-of-systems design spans a rather long period, and the needs, priorities and goals may change over time, it is important that the selected test strategy incorporate test process adaptation and continuous improvement. This means adaptation to constraints and developments, continuous improvement to increase effectiveness and efficiency of the testing and quality processes implemented.

If processes are automated, it will be necessary – periodically – to ensure their efficiency and to measure them periodically (e.g. with fault injection techniques).

3.3.2. *Important points in the strategy*

In addition to adapting to and fitting properly into your development cycle, your test strategy should ensure that the sequence of activities within various test levels – and multiple test environments – demonstrates that all types of defects have been researched (and all defects potentially found) before the component, product or system is moved to the next stage. You must therefore ensure that:

– At the start of a phase, the quality level of the input deliverables (specifications, data, code, etc.) is of sufficient quality. This involves two things: (a) ensuring – through checks carried out at the start of the phase – that the expected quality is present (if this quality is not present, it will be necessary to refuse the deliverable); (b) perform checks at the end of the phase to ensure that the planned and expected quality is present. This may involve looping through the phase until the quality level is achieved.

– The test phases complete each other correctly without doing the same work twice. It is inefficient to perform tests with the same objective (finding the same anomalies on the same objects to be tested). Each of the test tasks – if there are several in the same test level – having the purpose of finding the defects of a type, it is necessary to ensure their execution.

– Test tasks provide realistic metrics and KPIs, proving that they were executed correctly and that previous tasks were too. This activity of controlling the execution of tasks makes it possible to ensure that the work has been carried out correctly as well as the targets for continuous improvement have been reached. In case of inconsistent measurements, the test manager should investigate the causes and implement improvement actions.

– Risk levels go down and the number of risks decreases. This involves periodic risk monitoring, for example, at the start of each iteration or at least during each PI planning, and the implementation of risk mitigation actions.

– The tasks and their objectives are measurable and achievable. These tasks and objectives can be considered as requirements and therefore must be SMART.

– The strategy be understandable by each of the stakeholders so as to have an understanding of the ins and outs, as well as everyone's support on the importance of every task mentioned in this strategy.

Finally, you will have to make sure that the strategy adapts to the chosen development cycle; for example, in an Agile environment you could do it by:

– performing system tests at the end of each iteration;

– providing for one or two stabilization sprints at the end of iterations to fix flaws without introducing new features;

– performing operational, technical (performance and security) and user acceptance testing before delivery to production.

If you work in a DevOps environment, some of these tasks will be automated to remain in continuous delivery.

3.3.3. *Strategy monitoring*

To be effective, a strategy must be tracked throughout the project and may need to be adapted if necessary. This involves quickly identifying deviations by taking measures and complete and consistent reporting, and then proposing corrective actions that will allow the initial objectives of this strategy to be achieved.

The audit should focus on the quality of the processes, that is, their efficiency and effectiveness. Efficiency can be measured by the workload divided by the deliverables (e.g. J/H per function point, effort per user story, etc.). Efficiency can be measured by the FTR (first time right) metric, that is, the number of deliverables delivered without any defects (see also section 16.3.1 of volume 2).

In the deliverables we include the specifications, descriptions of user stories, features or epics, as well as the code, the classes and test methods (in the case of TDD), the specifications of unit tests, integration, end6to6end, system, acceptance, etc., the documentation, as well as any information produced or used during the realization of the system-of-systems set up by the test strategy.

Continuous verification should include dashboards that periodically measure and display information relating to deliverables and processes, allowing results to be compared over time.

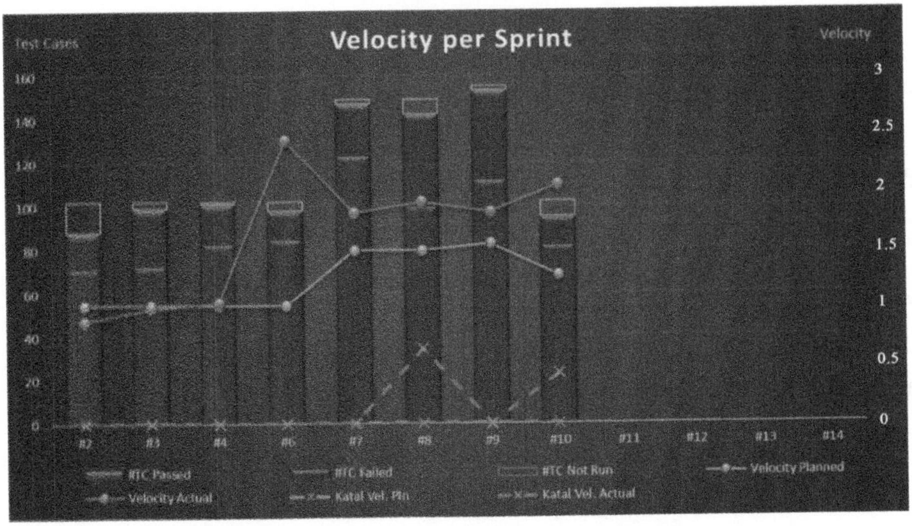

Figure 3.2. *Test execution velocity per sprint. For a color version of this figure, see www.iste.co.uk/homès/systems1.zip*

We see in the graph of Figure 3.2 that the planned velocity (velocity planned) is exceeded by the current velocity (velocity actual), but that the percentage of test cases not executed or in error remains very high during the sprints. The conclusions that can be taken from this reporting are, among others, that (1) it may be necessary to reassess the number of TCs that can be produced/executed per time period, in order to have a planning closer to reality; (2) if the planning is correct, it may be necessary to check the quality of the elements produced, to ensure that no activity has been overlooked; (3) sprint #6 had a significantly greater increase in velocity than all other sprints, but was it at the expense of execution quality or intended coverage?

Without continuous verification of each of the build and test processes in a system-of-systems test strategy, it is unlikely that the desired result will be achieved.

3.3.4. *Shift left, costs and time*

Whether the development model is V-cycle or Agile development (Scrum, NEXUS or SAFe), the goal is to find flaws in the system-of-systems as early as possible to reduce costs and lead times, at a minimum.

The “shift-left” attitude aims to find defects as early as possible because their correction costs less at that time.

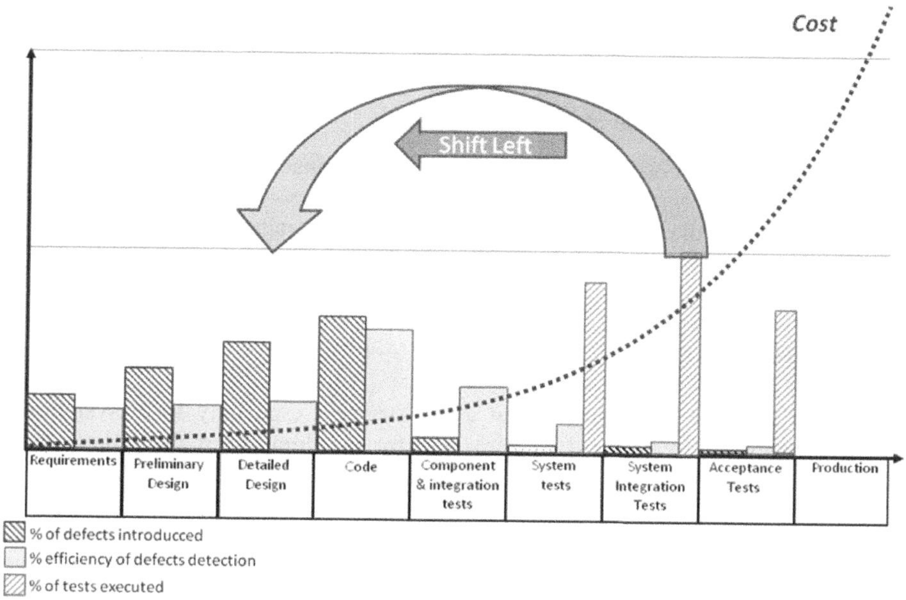

Figure 3.3. *Shift left*

Anomalies are mainly introduced in the requirements and design phases, up to coding. Other anomalies – regressions – can be introduced in the later phases, but in general, their number is generally lower.

Testing early reduces defect-correction costs because the costs of detecting and correcting anomalies increase over time. This increase in costs comes from the fact that prevention is better than a cure. Identifying and correcting a defect early means there is no need to also fix test cases, nor rerun test cases, nor provide new fixes (involving regression testing, etc.).

We see that the most frequent testing phases currently (system, system integration and acceptance testing) are performed at a time when the cost of remediation is high. In addition, these corrective activities take time and come at a time when there is not much time left before going live: it would be useful to test earlier.

According to the work of Capers Jones (2007) and Jones and Bonsignour (2012), the efficiency of fault detection in the early phases is greater than in the post-code phases. It is more cost-effective to detect defects as early as possible:

– on the one hand, because it makes it possible to focus on the typologies of defects introduced at each of the phases;

– on the other hand, to save time by correcting early, and avoiding the introduction of defects at the input of later phases;

– finally to reduce costs by reducing the financial impact of corrections.

We clearly see the interest of this "shift left" with regard to the detection of defects.

3.3.5. *"Optimal" strategy*

The test manager must realize that a test strategy, especially for a system-of-systems, cannot be created and remain fixed for the duration of the project. It is therefore necessary to design an optimal and modifiable strategy.

To be "optimal" in terms of effectiveness – finding all faults – and efficiency – finding faults with the least possible effort and cost, a strategy must be simple, understandable by the greatest number of people and adapted to the needs of impacted organizations.

The "optimal" test strategy must make it possible to achieve – with limited resources – the objectives defined in the organization's test policy – for the product or system-of-systems under test – and the objectives defined for this system-of-systems, including about its dissemination and use. This implies that the analysis, design, implementation, test execution and test reporting activities are adapted to the test aspects to be covered – functional and technical – and to the level of quality to be achieved. These test conditions may relate to technical elements or business elements, or to commercial elements such as respecting a date of placing on the market.

Different test strategies and methods can be applied to the different test levels and to the various elements of the system-of-systems. A risk-based and requirement-based analytical strategy, complemented by pattern-based testing, advisory strategies and a dose of reactive testing can limit the negative elements of each of these strategies and maximize their benefits.

Whatever the methods and strategies considered, it is necessary to anticipate the effectiveness of each of the tasks, to determine whether the goals can be achieved with the resources available, then to measure the effectiveness actually achieved by the tasks. This amounts to determining, for each activity, quantified objectives that will serve as a benchmark for monitoring the progress of these activities. It is an illusion to think that all defects identifiable on level will be identified by a single test campaign and corrected correctly. Therefore, an optimal strategy must anticipate both the rate of discovery of defects by test execution, the rate of correction of these defects and the rate of injection of new defects following these corrections.

The effectiveness, efficiency and satisfaction metrics periodically measure the progress of the level of success about these objectives.

An “optimal” test strategy can only be for a specific system-of-systems or software. The context, the objectives, the success criteria will be different from one environment to another, from one organization to another. It is therefore critical (for the test manager in charge of designing the test strategy for a system-of-systems) to identify the particularities (risks, challenges, etc.) of their project. Next, it will be necessary to define what constitutes a quality system, subsystem or component for the system-of-systems and how to measure this level of quality. The choice of a quality standard (e.g. ISO25010) ensures that all quality characteristics are taken into account in the design of the strategy. It should not be limited only to functional needs.

3.3.6. *Ensuring success*

Once the test strategy has been chosen, it will be necessary to describe its organization, the objectives for each test level, the methods to be used and the means of measuring progress (KPI). This description must be sufficiently detailed to be understandable and acceptable to the stakeholders. This is done by writing a master test plan and level test plans.

To ensure that the expected level of success is achieved, it is necessary to be able to link the results obtained (the test results) to the expected results (the defined requirements). This involves defining what will be considered a success – for the test level and for the system-of-systems – in a measurable way, implementing metrics to report the measures, analyzing the reported measures and follow trends and progress.

It will be necessary to ensure that the hypotheses taken at the start of the project are verified – to compare the expected and predicted values with the actual values resulting from the execution of the tests – and, if this is not the case, it will be necessary to propose and implement appropriate mitigation solutions.

Ensuring success can be complemented by measures to verify specifications (via reviews) or follow-up on deliveries (via “health checks” or NRTs performed in the production environment.

3.3.7. *Why multiple test iterations?*

One of the questions to ask is the minimum number of test iterations to achieve our anomaly detection objectives. Some organizations consider that three iterations are sufficient; our experience shows us that this is a utopian vision. Let us take an example: consider that we use several complementary test techniques in order to achieve a detection rate of 50% per test iteration; a 50% limitation due in part to the impossibility of running all the tests on the software (e.g. due to blocking anomalies).

During the first iteration, we will therefore find 50% of the anomalies, then 25% more (i.e. 75% of the total number of anomalies initially present in the delivered software) during the second iteration. Subsequent iterations will increase the total to 87.5%, 93.75%, 96.87% anomalies, and so on. This will result in an anomaly detection curve similar to those proposed by Weibull.

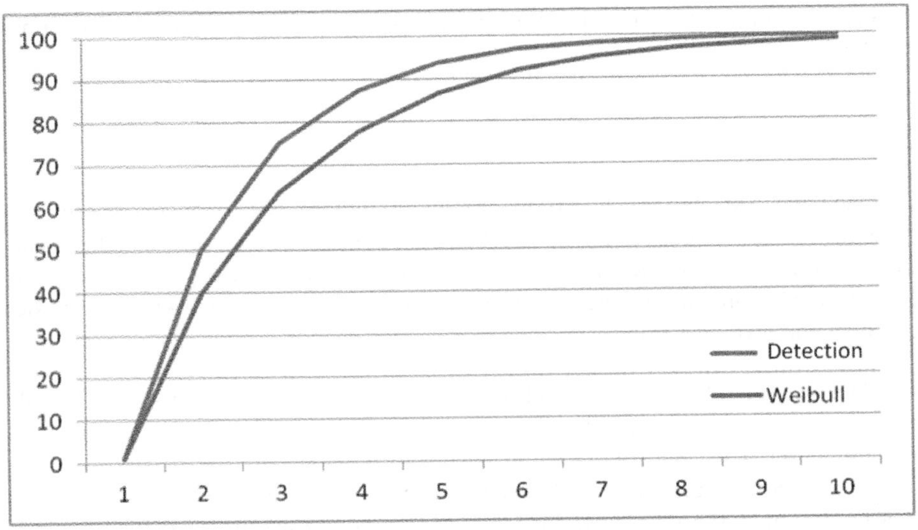

Figure 3.4. *Necessary test iterations. For a color version of this figure, see www.iste.co.uk/homès/systems1.zip*

Our objective (95% detection) is only reached from the fifth iteration (96.87%) and if we had set it at 99%, it would have been necessary to plan at least seven iterations. The curves are in percentages, and if the total number of defects in the application is large – as input to the testing activities of an iteration – the number of anomalies delivered to users will be higher. In our initial example (300 defects), after five iterations, nearly 10 anomalies will be delivered, while we will need nine iterations to drop to one delivered anomaly. The lessons of these curves must therefore be related to the total number – estimated – of anomalies in the software.

However, this calculation is subject to a few assumptions:

– The first hypothesis is that each new iteration will find 50% of the remaining defects. This implies that the techniques implemented previously did not make it possible to identify these defects, or that the level of coverage of the application did not make it possible to identify these defects.

– The second hypothesis is that the software is at an advanced level of maturity when delivered to the test team, and that the developers do not add new functionalities (and therefore new defects) during the correction of previously identified defects. If so (addition of other features during bug fixing), we should have increased the total number of bugs, and increased the number of iterations needed to reach our quality goals.

– The third hypothesis is that the duration of the test during each iteration is similar, since the workload is similar. Indeed, it is necessary to verify that everything that worked previously continues to work (regression tests), that everything that has been identified as an anomaly works during this iteration (retest or confirmation test) and that what does not has not yet been tested is indeed. If we don't test all the software – for example, by ignoring what worked before – we make a risky assumption, based on blind faith in the quality of our developers, and their ability to avoid making mistakes when fixes. But it is the same ones who introduced the defects in the initial code.

– The fourth assumption is that the defects are corrected at the end of the iteration and do not reappear later. This implies several things:

- a correct configuration management avoiding the propagation of an anomaly between several iterations;

- the capacity of the development teams to deal with all the anomalies during the duration of the iteration, without introducing delays in the process;

- that the deliveries of patches within the iterations be limited, so as not to introduce delays by the obligation to reinstall and reset the test environment (software, data, etc.), to carry out retesting and regression testing, as this will impact workload – and timelines – without improving the quality of the software.

– The fifth assumption is that the treatment of anomalies does not take significant time during an iteration. This is obviously false, the anomaly management activities (e.g. drafting of anomaly sheets, retest and regression tests (NRT), and other tasks of monitoring the evolution of anomalies, plus all the development tasks correctives) taking all the more time as there are anomalies identified during an iteration. The size and complexity of the software also has an impact on the number of anomalies and the workload required to manage them.

To think that we can arbitrarily reduce the duration or the number of iterations of the tests and keep the same level of quality is illusory: the efficiency of the testers and the capacity of correction of the developers are not extensible. Developments often (always?) take longer than initially planned, and delivery dates to customers are less flexible than those to test teams. Consequently, a reduction in the durations and the number of test iterations will necessarily be counterproductive. The curves in Figure 3.4 perfectly illustrate the impossibility of reducing the tests without reducing the level of quality of the delivered software.

Increasing the rate of anomaly detection by testers is possible and requires training, time to set up these trainings, and time for them to practice and reach a level of competence in their use.

It is important to realize that increasing the anomaly detection rate is interesting, but much less effective if the number of anomalies is large: the number of residual defects will be higher. So, a reduction in the number of defects added during the design phases is more interesting and more profitable. This can be done through static analysis tools, reviews and inspections of specifications, requirements, codes and architecture.

3.3.8. *Progress forecast*

Choosing a testing strategy and methods does not guarantee that the implementation will be correct or that the expected success will be achieved. To ensure the success of the planned activities, it is also necessary to ensure that each activity has been correctly planned (definition of the profiles, durations and loads of each activity in line with the objectives defined for this activity, in terms of loads used, deliverables generated, etc.) to have a baseline from which to compare the actual progress.

This implies:

– when defining strategy activities, define objectives in terms of workload, deliverables, deadlines and level of quality; to ensure that these objectives are measurable, by defining metrics;

– identifying prerequisites and dependencies of each activity, to ensure that the delivery milestones will be met;

– informing and anticipating the impacts of projects and activities on each other;

– during the execution of activities, periodically measure whether the achievements are in line with the anticipated objectives.

It will therefore be necessary to have relevant progress data for each activity and statistical information to measure the overall progress of the strategy.

This measurement of progress is made through the design of metrics (KPI: Key Process Indicators) which will be used to compare actual progress with forecasts.

NOTE.– as Nils Bohr said, “it is very difficult to make predictions, especially regarding the future”. You will be led to make estimates which will never be exact, but will be useful to you by indicating tendencies. It can be useful to either (1) explain orally, during presentations, the margin of uncertainty linked to these forecasts or (2) use the functionalities of spreadsheets to show these uncertainties – positive and negative – in relation to a curve.

3.3.9. *Continuous improvements*

To think that a test strategy can be directly and eternally applicable is illusory. Changes will take place, within or outside the organization, and they will influence the way of developing or testing, the levels and durations of test effort, even the objectives and quality criteria on which to focus. Adaptations will therefore be inevitable.

It will be necessary to measure the level of performance of its test processes, to consider adapting them in order to make them even more efficient and to define improvement objectives. From there, it will be necessary to implement these changes, then measure if the objectives are met and repeat this periodically. This continuous improvement is standard and included in the concepts of agility and DevOps. It is also a good practice for sequential developments. Continuous improvement should be included in any test policy and declined in the test strategy of each application.

To effectively implement a continuous improvement policy, it is necessary to continuously measure the current level of performance and identify the trends that are emerging over time. It is necessary to identify if these trends are not going in the desired direction; it will be necessary to implement corrective actions – or consider changing the objectives – to be able to achieve them.

4

Testing Methodologies

Test policies and strategies will rely on the implementation of test methodologies to achieve their objectives. Several software application testing methodologies exist. A test methodology describes the method that will be used to execute the test activities in a certain order in order to achieve the expected objectives: an application – or a system-of-systems – that achieves its objectives. Note that we do not say "defined in the requirements" because some requirements may be implicit or implied.

The test methodology must be able to explain the activities to be performed and their scheduling, with enough clarity to remove ambiguities and identify incoming and outgoing deliverables. The methodology will be developed in the Test Plan and/or in the Test Strategy.

4.1. Risk-based tests (RBT)

The principle of the risk-based testing (RBT) method is to identify and prioritize – during risk analysis – product risks. These risks will then be dealt with in decreasing order of priority, so that if the project has to be interrupted before all the identified risks have been dealt with, only the lower priority risks will remain to be dealt with.

Risk analyses will also identify "project" risks, that is, instances where the test project itself is subject to risk. These risks, just like the inter-project dependencies, should be anticipated in order to be able to plan mitigation activities, to reduce the impact or the probability of occurrence of the risk.

The risk-based testing strategy influences testing effort by increasing testing effort for risks considered significant and reducing – or even eliminating – testing effort for risks considered minor. The number and complexity of test cases will vary according to the criticality of the risks.

Further and more detailed analysis of risk aspects is available in Hubbard (2008) and in Black (2008).

4.1.1. *RBT hypothesis*

Applicable hypothesis are:

– the exhaustive assessment – and a prioritization – of the risks is carried out for the system-of-system and for each of the systems, subsystems or components that it comprises, from the start of the project and at each level. A periodic risk reassessment takes place;

– the components are developed by the production team and delivered to the test team in descending order of risk: first the components dealing with the most risky elements, then the least risky elements;

– if the duration of the project does not allow us to cover the tests of all the functional or non-functional characteristics, it is acceptable to ignore them. This implies that the contract must provide for the possibility of not delivering all the requirements provided for in the specifications;

– in systems-of-systems, risk analysis must be based on the business risks downstream of the changes made to the systems. These risks may not be immediately apparent and result from cascading impacts on other systems. This implies knowledge of all the elements composing the architecture of the system-of-systems, flows and volumes, the impacts of the decommissioning of applications, parallel processing of old and new applications (in the case of replacement according to a phasing spread over time), etc.

4.1.2. *RBT methodology*

Risks are one of the main criteria for prioritizing tests: after having identified the risks, we will sort them in descending order of priority and limit the highest risks before dealing with the lower risks. The risks apply initially:

– on the product and its ability to provide the services required by users, then we will speak of "product" risk;

– on the project and its ability to achieve its objectives, we will talk about "project" risk.

Requirements can be considered a risk of customer dissatisfaction if they are not met. We will therefore treat explicit requirements (e.g. contractual), implicit requirements (e.g. deduced from the architecture) and risks (e.g. forgotten requirements) in the same way.

To be able to prioritize the risks – and the requirements – it is necessary to have a common scale representing the level of risk. This level of risk is defined as the product of the impact and the probability, using a simplified scale for the impact and for the probability of occurrence of the risk (e.g. scale from 1 to 5 where 1 is the strongest and 5 the weakest). The result of the product (called RPN for Risk Priority Number) would therefore be a number ranging from 1 to 25, where a risk evaluated at 1 is both very impactful and very frequent, and a risk evaluated at 25 is both very low, impactful and very infrequent.

The RPN is used to determine the order of processing risks and the level of testing effort required for that risk. A very high risk (e.g. RPN between 1 and 5) will require a greater test effort than a low risk (e.g. several tests using several techniques), as proposed in Figure 4.1.

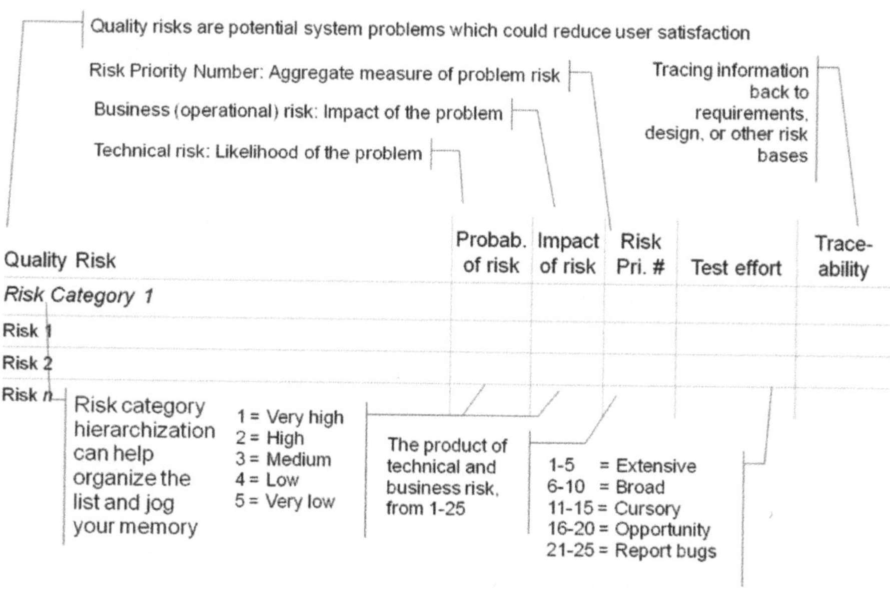

Figure 4.1. *Example of test effort based on RPN [© RBCS]*

4.1.2.1. *Test effort and risk level*

For a very high risk, it will be necessary to implement multiple testing methods and techniques, to increase the level of coverage and therefore increase the testing effort. This mechanically increases the cost of the tests and their duration.

An example of defining test actions according to the level of risk is the standard/DO178B/ED12B (EUROCAE, DO178 B - ED-12B). This standard, dedicated to software in civil aircraft (airplanes, helicopters, etc.), defines categories of failure conditions (failures):

– Catastrophic: failure conditions could prevent continued safe flight and landing.

– Dangerous: failure conditions likely to reduce the capabilities of the aircraft or the ability of the crew to deal with hostile conditions by up to:

1) a significant reduction in safety margins or functional capabilities,

2) physical problems or increase in workload such that the crew is no longer able to perform their task accurately or completely, or

3) negative effects on the occupants such as serious or life-threatening injuries for a small number of them.

– Major: failure conditions likely to reduce the capabilities of the aircraft or the ability of the crew to cope with hostile conditions of a severity resulting, for example, in a significant reduction in safety margins or functional capabilities, a significant increase in the workload of the crew or conditions affecting their efficiency, or discomfort for the occupants, possibly involving injuries.

– Minor: failure conditions that do not lead to a significant reduction in the safety of the aircraft, and that are likely to lead to actions for the crew that are entirely within the scope of their abilities. Minor failure conditions may include, for example, a slight reduction in safety margins or functional capabilities, a slight increase in crew workload such as routine flight plan changes, or some inconvenience to occupants.

– No effect: failure conditions not affecting aircraft operational capability or resulting in increased crew workload.

Then, the DO178B/ED12B standard describes the software levels and the test evidence to be provided to ensure that the levels are met:

– Level A: software whose malfunction would cause or contribute to a failure of a system function resulting in a "catastrophic" failure condition for the aircraft.

– Level B: software whose malfunction would cause or contribute to a failure of a system function resulting in a "dangerous" failure condition for the aircraft.

– Level C: software whose malfunction would cause or contribute to a failure of a system function leading to a "major" failure condition for the aircraft.

– Level D: software whose malfunction would cause or contribute to a failure of a system function resulting in a "minor" failure condition for the aircraft.

– Level E: software whose malfunction would cause or contribute to a failure of a system function leading to a failure condition with no effect on the operational capability of the aircraft or on the pilot's workload.

Software level	**A**	**B**	**C**	**D**
Compliance of high-level requirements with system specifications	**Oblig.**	**Oblig.**	Option	Option
Precise and consistent high-level requirements	**Oblig.**	**Oblig.**	Option	Option
Low-level requirements in accordance with high-level requirements	**Oblig.**	**Oblig.**	Option	
Test coverage of high-level requirements ensured.	**Oblig.**	**Oblig.**	Option	Option
Low-level test coverage is provided	**Oblig.**	Option		
Structural MC/DC coverage (modified condition/decision) provided	**Oblig.**			
Structural coverage (decision) assured	**Oblig.**	**Oblig.**		
Structural coverage (instruction) provided	**Oblig.**	**Oblig.**	Option	
Structural coverage (data coupling and control coupling) is ensured.	**Oblig.**	**Oblig.**	Option	

The table above shows that in cases of level A software, the burden of testing and proof of test execution is significantly greater than for levels B, C or D (there is no obligation of proof of test for level E). The test strategy imposed by the standard will have the impact of increasing the test effort for software where the criticality of the failure is high.

In the context of your projects, the criteria of DO178B are probably not applicable, especially if there is no risk of death or injury to individuals. On the other hand, the principle can be picked up with a modification of the definition of failures (e.g. loss of functionality without the possibility of bypass, loss with possible bypass, slowness, etc.), by providing:

– clear definitions of the criticality levels of failures, and from there the components (or software or parts of software) that can cause such failures;

– a list of actions to be implemented according to the failure criticality level.

The test strategy can be defined by test level, in addition to the level of criticality of the software component. This implies that each of the test levels will focus on one or more types of defects. The sum of all levels will represent the sum of all tests. The risk of such a strategy is that it assumes that the tests planned on each level will be correctly executed.

4.1.2.2. *Risks and dependencies*

Dependencies of components on each other – or even of one feature or user story on others – are a risk like any other. It is therefore important to develop the components in an order that respects the dependencies, to prevent the people in charge of the tests – and sometimes the production teams – from being blocked until the realization of the elements on which they depend. Similarly, an identification of dependencies will identify which components must be delivered and tested before the others.

Dependencies are not limited to software or system components but must also be identified in terms of activity execution prerequisites, test environments (including their representativeness), test data and resources for design and execute the tests. In some cases, too, an identification of dependencies will make it possible to manage activities correctly.

Component developments can take longer than expected and impact other components – and the test cases that verify them – generating risks of delays, dependencies and blocked test cases, etc. Similarly, specifications (e.g. features or user stories) may have to be modified after the design of the tests, which will lead to code modifications as well as the need to adapt the test cases, or even to redesign them, and to adapt the data used. It is therefore critical to correctly identify the functional scope of the tests to be carried out and the technical components that will be tested, and to ensure that the components to be tested are provided within the prescribed deadlines with the expected level of quality. Each delay or modification should be announced to the other parties in order to allow the schedules to be adjusted if necessary.

Prioritization of the components to be designed will depend both on their development and test duration, as well as on the number of components and their dependencies. This dependency must be considered for the prioritization of features/user stories (case of Agile developments) and for integration needs.

4.1.2.3. *Risk coverage*

Once the risks and their dependencies have been identified – to be considered as other risks – it is necessary to manage their evolution, which goes from appearance to mitigation to disappearance. As the number of risks is large, the test manager must ensure that each of the risks is identified, measured in terms of impact and frequency, and that impact and/or frequency reduction actions are identified (and executed).

Risk inventory: this recurring and periodic activity (e.g. weekly) allows you to identify new risks that appear and to remove risks that may no longer be relevant. Let's take an example of a risk that can disappear: if the identified risk is "the client could reject the architecture documents at milestone X", then, if milestone X has passed and the client has not made any remarks, the risk disappears.

Periodic risk assessment: this activity should cover all identified risks (not just the top n risks), to identify if the impact or frequency has changed. It will be checked whether the mitigation actions previously planned have been carried out and have borne fruit (i.e. a reduction in frequency or impact). RPN shall be recalculated and one or more mitigation actions may be proposed.

Level of coverage: depending on the level of risks indicated by the RPN, you can define the necessary test effort (see Figure 4.1) and implement traceability from this risk to the applicable test case(s).

Risk reduction: at the end of each risk assessment activity, you can compare the results obtained with the results of the previous assessment iteration. You will thus obtain a measure of the level of criticality of the risks. Figure 4.2 shows you an example of reporting on the evolution of the number of risks per category per month.

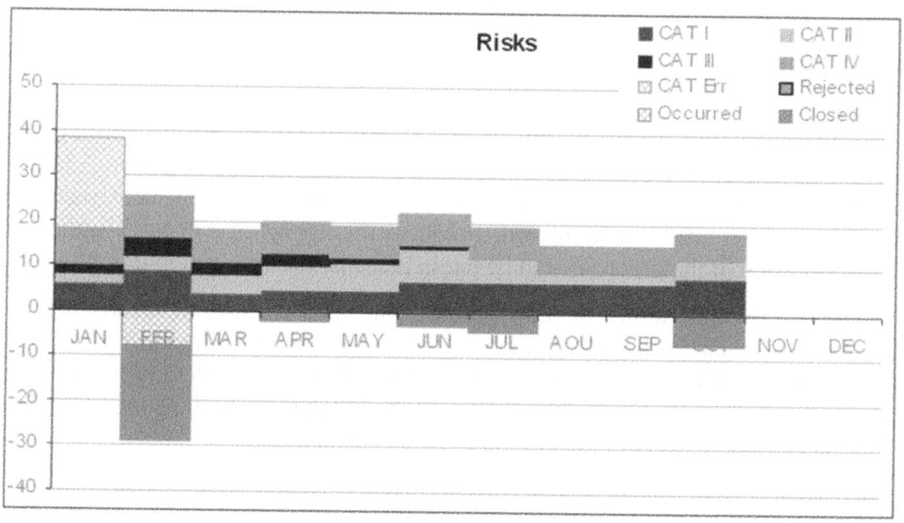

Figure 4.2. *Periodic risk assessment*

4.1.3. *RBT versus RRBT*

As mentioned earlier, RBT is a methodology where testing is risk-based. RRBT stands for risk- and requirement-based testing and implies that testing is both risk- and requirement-based. It is therefore a methodology merging RBT and TBX. With the RRTB methodology, it is difficult to prioritize requirements over risks. Should a risk with a very low probability, but a major or even disproportionate impact, be tested more than a requirement specified and requested by the business?

Combining risks (which generate negative impacts) and requirements (which generate positive impacts) can lead to a problematic assessment in that an unsatisfied requirement becomes a risk of customer dissatisfaction. This leads to a different assessment of the test cases to be implemented: should we confirm that the requirement is present or should we ensure that the risk is absent?

4.1.4. *Reactions to risks*

Several types of reaction appear when we manage risks:

– Ignore the risk: it is possible to consider that this risk will not appear in our environment, or that nothing can be done to limit the probability of its occurrence or its impact? This solution can be justified in case of low impact or probability. However, we must not be blind and consider that the risk has no reality.

– Transfer the risk: once a risk has been identified at a level (or on a system), it is possible to escalate the information to a higher hierarchical level and let them make the necessary decisions. This reaction may be justified if the investments are beyond what is acceptable or authorized at this level.

– Reduce the impact: the objective here is to plan and implement solutions that – when the risk occurs – will result in reducing the impact of the risk. For example, backing up data, scripts and test environments in order to implement recovery solutions on other sites is a solution that reduces the impact of a disaster in the test lab. A definition of the procedures to be put in place is often present in this case.

– Reduce the probability of the risk appearing: this consists of identifying the root causes of each risk and implementing actions to anticipate its occurrence and limit its impacts. This could consist of the analysis of metrics to identify whether the probability of the risk appearing increases and the implementation of preventive actions according to the results of the analysis of the metrics.

4.1.5. *Risk computation*

The level of risk is often defined as a product of impact and probability.

It is possible to determine the impact as a single factor or as a combination of three factors: the impact in time (deadline for completion) on the project, the impact in money (additional cost of completion) and the impact in performance (limitation of software capabilities). For each factor, we can determine value ranges, and combine these ranges to obtain a reference value for the impact. I advise using a range of values from 1 to 4 or 1 to 5, with 1 being the most impactful element and 5 representing the least impact.

Regarding the probability of occurrence of the risk, we can, here too, define ranges of values. I recommend a range of values from 1 to 4 or 1 to 5, with 1 representing the highest frequency and 4 (or 5) the lowest frequency.

Choosing a range of 1–4 allows you to link percentages to values in order to obtain 100% (10+20+30+40=100), while the range of 1–5 allows values such as very high, high, medium, low, very low.

The probability is not absolute: a one in a hundred chance may be an acceptable probability if the software is used once a year, and unacceptable if the software is critical or used every hour. In the case of hourly use, statistically there will be occurrence of the risk at least once a week (one week making 168 hours).

Identifying impact and likelihood may require actuarial studies which are not available to the test manager, so assessments may be wrong. It is for this reason that it is recommended that risk assessment be carried out jointly with representatives of customers, development and management.

4.1.5.1. *Measuring risks (1–4 values)*

As risks evolve, it is necessary to reassess them periodically. As a risk can have several impacts, it is necessary to assess each possible impact. We propose below an evaluation according to the probability of occurrence and the impacts on the product and on the project:

4.1.5.1.1. Probability of risk (1–4)

Condition	Probability	Estimation
The risk will almost certainly occur	90+%	1 – very high
The risk will not occur only on optimistic assumptions	60–90%	2 – high
The risk may or may not occur under normal circumstances; no clear evidence available for either possibility	30–60%	3 – average
The risk will only occur with pessimistic assumptions	0–30%	4 – low

4.1.5.1.2. Risk impact (1–4)

The severity of the impact is seen from three different angles that combine:

– additional costs induced by the risk;

– limitations or performance of the delivered product;

– impact on the delivery date of the product.

Cost	Performances	Planning	Severity
More than 25% additional costs Seriously outside budget	Inability to deliver the component, software or system within the acceptance criteria	More than 25% delay or inability to meet the commitments of the main system (the system-of-systems)	1 – critical
10–25% additional costs, well beyond budget	Inability to achieve a key criterion and no workaround identified	10–25% behind schedule or inability to meet one or more major milestones	2 – high
5–10% additional costs, limited budget excess	Inability to meet a criterion, but possibility of circumvention	5–10% late, or no need for major re-planning	3 – average
Less than 5% additional cost, within budget but not negligible	Inability to meet a criterion, but negotiable with the client	Less than 5% delay or delay can be contained	4 – low

4.1.5.1.3. Risk category (1–4)

Risks will be grouped into four categories handled by different hierarchical levels of responsibility:

– risk category I which includes values 1 and 2; this risk category is often managed at the level of company directors;

– risk category II which includes the values 3–6; this risk category is often dealt with by the person in charge of the system, or the system-of-systems;

– risk category III which includes the values 7–12; the risks of this category are often dealt with mainly at the level of those responsible for the subsystems;

– risk category IV which includes values 13–16; the risks of this last category are handled by the person responsible for the equipment or component in question.

This solution makes it possible to identify the risks to be dealt with according to their category (from the most important to the weakest) and what level of responsibility must deal with them. Periodic reassessment of risks can move a risk from one category to another.

4.1.5.1.4. Risk calculation (1–4)

The principle is to measure the severity according to the three axes (cost, performance and planning) and to define the overall severity. From this moment, the table below can be used to define the category of risks according to the probability.

	Probability			
	Very high	**High**	**Average**	**Low**
Critical - 1 critical or 2+ high	1	2	4	8
High - 1 high or 2+ average	3	5	6	10
Average - 1 average	7	9	11	14
Low - all low	12	13	15	16

4.1.5.2. *Measuring risks (1–5 values)*

An elegant solution initially proposed by Rex Black is to use a scale of 1–5 for the impact and the probability of occurrence. It is then easy to multiply the impact by the probability to obtain what is called an RPN (Risk Priority Number) varying from 1 to 25. From there, it is easy to define five categories of risks and associate with each category a level of test effort necessary to go from an identified level of risk to an acceptable level of risk.

4.1.5.2.1. Probability of risk (1–5)

Condition	Probability	Estimation
The risk will almost certainly occur	90+%	1 – very high
The risk will not materialize only on optimistic assumptions	71–90%	2 – high
The risk may or may not materialize under normal circumstances. No clear evidence available for either possibility	31–70%	3 – average
The risk will only materialize on pessimistic assumptions	11–30%	4 – low
The risk will only materialize under very pessimistic assumptions	0–10	5 – very low

You will note that the percentages represent a normal statistical distribution (bell curve).

4.1.5.2.2. Risk impact (1–5)

For the assessment of the impact of the risk, you can take inspiration from what has been defined previously or use the table below.

Risk impact	Estimation
Critical; impact on the survival of society	1 – very high
Major; impact on annual results	2 – high
Important; impact on quarterly results	3 – average
Minor; impact anticipated	4 – low
No object or very minimal	5 – very low

The risk can be assessed on the basis of various criteria: financial, product quality, delivery times, impact on the production effort, etc. In each case, the five impact categories must be defined from very high to very low (see Figure 4.1).

4.1.5.2.3. Risk categories and effort (1–5)

The product of probability and impact will generate a test load that will vary by risk category, with more testing effort for high-risk categories and less for low-risk categories. For example, we could have:

Cat. RPN	Effort	Test techniques to implement (example)
I 1–5	Extensive, full feature coverage with multiple tests per feature	EP, BVA, DT, path, exploratory
II 6–10	Broad, full feature coverage with one test per feature	EP, BVA, DT, path, exploratory
III 11–15	Superficial, partial coverage of most important features	Happy path, EP
IV 16–20	Opportunity: if we are brought into the functionality, it is tested; otherwise, it is not	Exploratory
V 21–25	Very small; Only escalate anomalies	None

4.1.5.2.4. Risk reduction (1–5)

Once risk assessment has been carried out,[1] it is necessary to define the actions to be implemented to reduce the risk to an acceptable level. You will find below an example of a solution that differentiates efforts to be implemented according to the initial level of risk and the critical functionalities (MC for Mission Critical) and not (non-MC). You will note that some aspects (entry criteria) relate to the design development activities (including test design), and that the risk mitigation measures also relate to the regression tests to be performed. The objective is to achieve an acceptable residual risk (here: low to very low).

Initial risk	Solution	Entry criteria	Coverage/mitigation	Exit criteria	Resulting risk
Very high	Premium	No process, unit test, review variance	MC: extensive non-MC: broad regr.: automated	High coverage, trust, quality	Low to very low
High	Gold	Few processes, unit test, review variance	MC: broad non-MC: superficial regr.: manual	Good coverage, confidence, quality	Low to very low
Medium	Silver	Some processes, unit test, review variance	MC: superficial non-MC: opportunity regr.: none	Adequate coverage, confidence, quality	Low to very low
Low	Basic	Discretionary process, unit test, review variance	MC: opportunity non-MC: none regr.: none	Low coverage, trust, quality	Low to very low
Very low	No test	–	No test	–	Very low

1 I would like to thank Rex Black for his inspiration and suggestion of this approach.

The premium approach requires that software development best practices and processes are followed, including good reviews of unit tests, requirements, designs, and codes. During system tests, critical functional parts (MC for Mission Critical) receive "extensive" testing, applications that are not "mission-critical" receive "large" testing and automated regression testing is performed. Exit criteria are defined in the test plan to support high coverage, high confidence and high-quality delivery. The level of residual risk expected upon delivery is low to very low.

The gold approach envisions the implementation of software development best practices and processes, including good reviews of unit tests, requirements, designs, and codes, but some minor variations are allowed. During system testing, mission-critical applications receive "broad" testing, non-mission-critical applications receive "shallow" testing and regression testing is manual. Exit criteria are defined in the test plan to support good test coverage, good confidence and good-quality delivery. The level of residual risk expected upon delivery is low to very low.

For the silver approach, all good software development practices and processes are recommended, including good reviews of unit tests, requirements, designs, and codes, but variations are allowed. During system testing, mission-critical applications receive "shallow" testing, non-mission-critical applications receive "opportunistic" testing and regression testing is omitted. Exit criteria are defined in the test plan to support adequate test coverage, adequate confidence level and adequate delivery quality. The level of residual risk expected upon delivery is low to very low.

For the basic approach, software development best practices and processes can be followed, including unit testing, requirements, designs, and code reviews, at the discretion of the project team. During system testing, mission-critical applications receive opportunistic testing, and applications that are not mission-critical will not be tested. Exit criteria are defined in the test plan to support low test coverage, low confidence and to allow delivery even if quality is low – after all, the project is a low-risk project. The level of residual risk expected upon delivery is low to very low.

The last level of risk (no tests) is considered not to warrant testing.

Note that this approach makes it possible to adapt the testing efforts to the levels of risks to consider, the categories of risks and the fact that the components – or functionalities – are critical for the application.

4.1.6. *RBT synthesis*

Identifying risks and setting up tests according to the level of risk identified is a simple and elegant way to allocate finite resources to an activity that requires – generally – much more load and time than what is available. This solution makes it possible to associate a higher test load with the greatest risks and a lower test load – or even zero – where the risks are considered to be lower.

It doesn't matter whether you choose a 4 or 5 level scale; you will need to use risk to drive both development and testing. It will be necessary that the developments of components considered "at risk" be carried out early, to allow more time for the execution of tests and the correction of anomalies. This may have an impact on the choice of components to be developed as a priority in the context of Agile developments.

Obviously, this depends on the identification and assessment of all risks.

Advantages	Disadvantages
Easy-to-understand method	Some organizations are reluctant to use the term "risk"
Applicable to product, project, short or long term	The acceptable level of risk is different depending on the individual, some people being more risk adverse than others
Allows us to quantify the costs and not to overinvest for low risks	

4.1.7. *Additional references*

As we have seen, the risk analysis to direct the tests focuses mainly on the calculation of averages. This methodology and these averages disregard the elements that are outside the various statistical curves. Many books such as Westerman (2007), Black (2008), Hubbard (2008), Taleb (2008), Tarantino (2008) and Savage (2012) identified potential shortcomings and omissions associated with risk analysis based on the calculations of averages.

4.2. Requirement-based tests (TBX)

The principle here is to define and prioritize the requirements and then process them in decreasing order of priority, from the highest priority requirements to the lowest priority requirements. In this book, we will use the term "requirement" to cover a requested need, or – according to Agile vocabularies – an “Epic”, “Feature” or “User Story”, even a “Use case”. The objective is to ensure that all contractual (functional), regulatory and technical requirements are developed, tested and delivered. If the requirements are prioritized, the assumption is that the realization will be carried out in the order of priority, from the highest priority to the lowest priority, while considering the dependencies. It is not usually planned to stop the project before all the requirements are delivered, but such a strategy can quickly provide significant added value.

The prioritization of requirements should influence the testing effort: a critical requirement will be tested more thoroughly than a minor requirement. The number of test cases associated with a requirement will depend on the level of prioritization of the requirement. In general, we will have the nominal test cases and one or more test cases covering the exceptions. It is recommended to have a statistical distribution of requirements that favors uncritical requirements.

The delivery of components, subsystems and systems to the test and integration teams is carried out according to the chosen integration methods, but must correspond – as far as possible – to the level of priority of the requirements (functional or not).

In Agile developments, requirements are often defined in the form of “User Stories” (see section 4.5). They are assumed to be unchanged during the iteration (the sprint) in which they are developed. It is important to note that in “Agile” developments, the choice of user stories to develop is often left to the development team, which can lead to delays in testing due to dependencies between user stories.

4.2.1. *TBX hypothesis*

Implementing requirement-based test methods involves the following hypothesis:

– Requirements are fully and clearly defined at the start of the project. In Agile-type developments, requirements are defined at the start of the project, but will be detailed at the start of each iteration. The choice of requirements to be delivered at the end of the iteration is left to the discretion of the implementation team.

– The requirements are stable throughout the project – in the context of sequential development – or throughout the iteration – in the case of Agile development.

In most business environments, requirements change over time. This is to some extent in contradiction with the assumption above. In sequential development, any change in requirements should result in changes to test cases and test procedures, and require new test executions. Configuration management helps us to manage these aspects properly. In an Agile development model, changes occurring after the first release should be treated as if it were sequential development. Changes that appear during the sprint or iteration should be pushed out of that sprint if following Agile orthodoxy.

4.2.2. *TBX methodology*

The design-to-requirement methodology is similar to the RBT methodology except that we will deal with the requirements and not the risks. Some requirements may be implicit rather than explicit. Very frequently, the defined requirements are all considered to have a high level of importance or criticality.

It is possible to link requirements and risks, considering that an unmet requirement is a risk of customer dissatisfaction. The TBX methodology thus joins the RBT method.

4.2.3. *TBX calculation*

The calculation of the level of advancement of the coverage of the requirements is simple but comes up against the level of granularity of the requirements. High-level requirement may be split into many, many lower-level requirements. To calculate a level of advancement, we will measure the number of requirements covered which we will divide by the total number of requirements. It is possible to group requirements according to their priority, to verify that the highest priority requirements are developed, tested and verified the earliest on the project.

4.2.3.1. *Prioritization propositions*

In requirement-based test and development, prioritization of requirements is based on the business value provided by each requirement. This financial value can be difficult to quantify, and the prioritization will depend on individual feelings, and often almost all requirements will be considered high priority or critical. You can consider a variety of ways to prioritize requirements:

– a solution to distribute the requirements would be to consider that a maximum of 10% of requirements would be critical, 20% major, 30% important and 40% minor to allow a total of 100%;

– another solution is to identify the aspects of business importance on a scale of 1–5 and complexity of realization on a scale of 1–5 (with 1 being the most important and 5 the least important) and consider the product (varying from 1 to 25) as the priority level. This solution integrates well with risk management and computation of RPN;

– another solution, suggested by Agile methods, is to let the development teams decide on the prioritization (Fibonacci suite, Planning Poker, etc.) and select requirements based on the team ability to finish in the allocated iteration time. This implies that the technical teams take precedence over the business needs of the company.

4.2.3.2. *Workload versus business impact*

The business teams mainly focus on the functionalities of the system. This focus will depend on everyone's vision, and therefore can take quite a bit of time and depend on the respective power of the various business representatives. This implies that technical constraints and quality aspects other than functional (i.e. product quality characteristic) have an impact and must also be taken into account.

Two other aspects to consider are:

– The refactoring workload, that is, technical improvements that do not represent new features but provide higher-quality code, at the expense of additional effort and potential defects.

– The regression testing workload to ensure the proper functioning of previously delivered features.

The prioritization of requirements and their inclusion – or not – in a development phase (sprint, release, etc.) is not always an easy task, but is critical for the prioritization of test activities and the definition of the testing strategy.

4.2.4. *TBX synthesis*

Requirements are the basis of all software development: without requirements, there is no reason to carry out any development! However, the evaluation of the criticality, the importance of each requirement compared to all the other requirements, can be problematic.

Advantages	Disadvantages
Simplistic	Requirements are often all considered very important or critical. This makes their ordering and prioritization difficult
	Requires early identification of all requirements and limiting changes to those requirements during the project
	Limits the ability to adapt the project to evolving business needs

4.3. Standard-based (TBS) and systematic tests

Testing based on standards (TBS) is a way of relying on good practices recommended by national or international bodies. Standards are constantly evolving, and some standards are replaced by newer ones (e.g. ISO9126 replaced by ISO25010, IEEE829 replaced by ISO29119, etc.) or newer versions (DO178B replaced by DO178B/ED12C).

As mentioned in Craig and Jaskiel (2007), systematic tests were born in 1985, based on the IEEE829-1983 standard. These are therefore standard-based tests. The work on IEEE829, the adaptation of its recommendations to the use of ISO29119 is quite simple.

4.3.1. *TBS hypothesis*

The principle assumes that the standards are the best industry practices, and therefore that applying them is a very effective solution, and it is true that the standards recommend practices applicable in certain cases, for example, in the aeronautics, space, medical, etc.

Standards are mainly applicable to security-critical areas and define test levels and evidence to be provided to guarantee the quality of the software.

The standards must be interpreted and understood by each of the parties active in the system-of-systems, to guarantee the consistency of the tests. Such an understanding can be problematic in the presence of subcontractors or co-contractors who could each have a different interpretation.

4.3.2. *TBS calculation*

There is no specific way of calculating how to measure and represent the coverage achieved by standard-based tests.

You can define a checklist, including each element of the standard that you consider important for your project, and check the presence or absence of each of these elements in all the deliverables where it applies. You will check the coverage rate of the standard. To check the coverage of the code by a standard, you will have to check – for each requirement, each US (User Story) and each component – all the elements of the standard that apply to it and then the satisfaction rate to be achieved and the rate achieved. It is therefore a very heavy workload to be applied only in specific fields (e.g. aeronautics, space, medical, nuclear, military, etc.) where the standards are applicable.

4.3.3. *TBS synthesis*

Standard-based testing is necessary if the domain of the system-of-systems to be tested is subject to certifications, approvals or other regulatory aspects that make the use of standards mandatory. The TBS can serve as a source of inspiration to increase the coverage of your testing strategies.

Advantages	Disadvantages
Offers an easy-to-implement basis for initial reflection	Is the standard applicable to your organization, to the domain of the system-of-systems and/or software to be tested
Mandatory in the case of systems-of-systems subject to standards (aeronautics, space, medical, nuclear, etc.)	Obligation to design one or more checklists before being able to measure
The techniques proposed in the standards and associated works allow you to get started and start testing efficiently.	Have the requirements been identified early and are their modifications during the project limited?

4.4. Model-based testing (MBT)

Model-based testing is an advanced technique that designs procedures, scenarios and test cases based on models of system requirements and behavior (Jorgensen 2009). The ISTQB (2015) adds that models must be formalized and contain enough information to generate test cases directly from the model.

The MBT method should not be confused with the representation – the modeling – of the behavior of the application that testers make when they design the conditions for testing software. The definition proposed by Paul Jorgensen defines a theoretical framework and proposes several types of modeling from which the scenarios and test cases to be considered can be deduced.

According to Utting and Legeard (2007) and the ISTQB, MBT must be supported by tools (e.g. Spec Explorer in VisualStudio from Microsoft). The model defines – with a formal language – the behavior of the software, which makes it possible to design the procedures and test cases (and the test data), to execute them and to follow their execution.

The basic idea is to improve the quality and efficiency of test design and implementation activities by:

– the design of a complete MBT model, typically with specific tools, based on the test objectives of the project;

– the use of this model as a specification to design the tests. The model must include a significant degree of formalization to allow the easy generation of scenarios and test cases.

Modeling increases the effectiveness and efficiency of testing by:

– increasing communication between stakeholders and facilitating the involvement of business analysts;

– creating a common understanding of requirements in the defined domain and identifying misunderstandings and ambiguities;

– avoiding defects in the early stages of development by validating the proposed model;

– reusing objects from previous projects;

– allowing the generation of different test suites (or sequences of test cases) from the same model, but covering different test objectives and/or levels.

4.4.1. *MBT hypothesis*

Like any test method, the MBT is associated with a number of assumptions that may prove to be inaccurate when testing a system-of-systems or a component of a system-of-systems.

The MBT does not replace traditional test design techniques, but allows testers to have a shared understanding of the domain rather than having – for each tester – a potentially erroneous individual view.

The models designed and shared can be incorrect, just as each tester can make mistakes when designing their tests. However, an erroneous model will have an impact on all testers, and even on developers. As the model serves as the basis for an automatic generation of test cases (in the case of MBTs supported by a tool), any evolution of the model will lead to an evolution of the test cases, often the generation of a new exhaustive series of test cases, test. The test team will then be forced to manage test versions (according to the model versions), and the test execution progress will have to be completely reviewed.

Black Swans.

As mentioned in Taleb (2008), any modeling generates a simplification of reality in the description of the model. So, the exceptional cases – the "Black Swans" described by Taleb – will not be included in this model and will not be tested.

This is valid whether the model is generated by the design teams or by the test teams, and whether the modeling is used during design – Agile or traditional – or during testing.

Like any tooled method, it should be kept in mind that the test tools, including the tools used in the context of MBT, may contain defects. It is risky to have blind faith in any tool.

4.4.2. *MBT calculation*

From the model, identify the combinations to be tested. If necessary, define an order of priority or criticality. Design the test cases to cover each desired combination and to cover the selected combination. Execute the planned test cases and measure, for each component modeled and/or for the complete system-of-systems, the percentage of test cases passed, in error and blocked.

4.4.3. *MBT synthesis*

Advantages	Disadvantages
Automatic generation of procedures, scenarios and test cases	The use of modeling elements defined by the developers is to be avoided.
Easy automation of tests and their execution	Need for testers to learn a new – formal – language
Save time running tests	Complexity of the models or obligation to design several models each covering different levels of complexity.
Automated traceability	On a system-of-systems, the modeling will have to reproduce the whole system-of-systems, which is a double development
Enables end-to-end test design through business process modeling	Are the defects identified due to an incorrect model or incorrect development
Enables "correct by design" solutions, as long as all requirements are known and proof is provided	A combinatorial explosion of test cases can take place depending on the tools used for the generation of test cases
	Ignores unmodeled exceptional combinations (black swans)

4.5. Testing in Agile methodologies

The so-called "Agile" methodologies are intended to respect the agile principles defined in the Agile Manifesto (see section 2.3.1) and revolve around three main methods: XP (eXtreme Programming), Scrum and KANBAN. Agile methods are mainly applied to small projects or applications and are based on the application of a set of good practices.

The principle of Agile developments is to develop first what has the most value. In a system-of-systems that we speak of technical value or business value.

One of the main difficulties in applying Agile methodologies in a systems-of-systems environment stems from the wide variety of systems, development methods and design teams. It is necessary to find solutions to integrate "agile" developments in environments involving a single delivery, or a phasing of deployments over a fairly long period.

4.5.1. *Agile "test" methodologies?*

In the Agile world, we mainly talk about TDD, ATDD and BDD. These techniques are a way of automatically verifying that unit specifications have been correctly developed. However, these are not ways to independently verify and validate that requirements or risks are properly covered.

The "TDD" acronym for "test-driven development" implies the design of tests in code and their execution with each build. This solution ensures the rapid identification of regressions in the code.

The acronym "ATDD" for "acceptance test-driven development" relates to the acceptance of the code for its integration into the system under development, and not to the acceptance of the system (or the system-of-systems) by the end users. Forgetting this distinction exposes you to misunderstandings between development teams and test teams.

4.5.2. *Test coverage*

In the Agile world as in the world of sequential development, it is important to ensure that the tests correctly cover both the desired functionalities (the requirements, user stories, features and epics) and the code (e.g. ensure that there is no dead or useless code).

Each team member is responsible for the quality of the deliverables they produce. This goes beyond just the development and testing team, and involve PO (Product Owners) and SM (Scrum Master), as well as business representatives. It is therefore necessary to define how to measure and monitor the level of coverage, from the design of the desired functionalities to the components produced.

In non-object-oriented environments, functions, modules and scripts can each be considered unitary components. On the other hand, in object-oriented environments, the smallest elements that can be individually tested are the classes and the methods associated with them.

The following sections are inspired by Linz (2014), where you can find more information.

4.5.2.1. *Classes, objects and methods*

Classes are abstract constructs composed of variables (attributes or properties) and methods (or functions) that act on these variables and their content. It is only during its execution that the class creates the objects (instances) which will exist in the memory of the system where they will be executed. Multiple instances of the same class can exist simultaneously.

It will therefore be necessary to cover each of the classes and each of the methods of these classes, considering the equivalence partitions (valid or not) and limit values associated with the variables. Variables can be private (exist only in the created instance) or public (visible from outside the instance) and therefore be influenced by actions of other components, classes or methods. The combinatorics quickly become enormous, and one solution is to use methods such as TDD and the test classes that this implements. Some programming languages facilitate this design of tests before designing the code.

4.5.2.2. *Object states and state transitions*

As in the object-oriented world, objects can have various states. Depending on the state, the possible actions (implemented by methods) may vary. State testing and state transition approaches should be used to identify:

– missing transitions between states;

– unauthorized transitions (method accepted when it should be rejected);

– erroneous actions such as methods producing incorrect result;

– incorrect resulting states such as what can happen with an incorrect variable value resulting in a random or unspecified arrival state;

– the use of state and transition tables is very useful in order to identify the combination of states and transitions – valid and invalid – to be tested.

4.5.2.3. *Unit tests*

In the world of agility and often in sequential development cycles, unit testing – sometimes called component testing – is the responsibility of developers. These are often limited to verifying that the code is working properly rather than checking that it is not malfunctioning (and missing out on defects). Developers suffer from author bias – in that they will not check something they have not developed – and often do not have the training – or the mindset – to design good tests that cover all probable causes of defects.

It is at the level of unit tests that we can effectively implement static analysis activities such as code analysis (with static analysis tools) or reviews (reviews and inspections of requirements and/or code), which make it possible to analyse the code without executing it (thus before it is finished) and to identify faults very early.

If we wish to perform dynamic tests at the unit or component level, to verify their operation, it is often necessary to design drivers and stubs to provide input data to the components and to analyse resulting data.

The following fundamental testing techniques (dynamic testing) should be implemented to verify component behaviors:

– valid and invalid equivalence partitions (EP);

– analysis of the boundary values of valid partition (BVA);

– combinatorial tests such as decision tables (DT); and

– state and transition tests (STT).

It is also during unit testing – and during code design – that techniques such as TDD (Beck 2003) can be used.

Ensuring comprehensive unit test coverage helps find defects sooner and easier. Unfortunately, unit testing is too frequently limited to ensuring that the software does what it is supposed to do, instead of ensuring it is not doing what it is not supposed to do. Thus, it becomes necessary to test more thoroughly at later stages of development.

4.5.2.4. *Integration testing*

When dealing with a single piece of software, integration testing focuses on integrating various components into a single system. In the framework of systems-of-systems, we must consider both the integration of software components into a single software system, as well as the integration of components, products and systems with the other components, products and systems that make up the system-of-systems. The complexity and difficulty are exacerbated and it will be necessary to take into account:

– Test environments that vary according to the level of integration. In the case of software integration tests of a software component, the environment is limited, but for the integration of products, including hardware and software, it will also be necessary to take an interest in the connectors (impedances, cable length, shielding), electrical/electronic protection of connectors, overvoltage protection (e.g. to be resistant to EMP-type pulses) and optical or radio connectors (WiFi, Bluetooth, etc.) which will involve designing test benches as representative as possible.

– The fact that it is the first moment when components, products or systems are integrated with other components, products or systems, coming from partners, suppliers or co-contractors.

– The versions of each of the components, products and systems, so as to identify the valid or invalid combinations. This implies that the installation of a modified version of software on the same hardware component could impact the product and therefore its reference in configuration management.

The automation of these tests is recommended, as far as possible; the integration tests to be sure to cover by tests all the exchanges, messages and interfaces between all the components makes it possible to anticipate rare combinations which can be problematic. Complete integration test coverage – that is, all messages, all data exchanged between components – is necessary whenever a component, product or system evolves. The evolution of the data passing through these media may also require the execution of an integration test campaign.

4.5.2.5. *Continuous integration*

Continuous integration consists of continuously integrating the changes made to the computer code of a project, to immediately correct any errors. Continuous integration requires test automation on a specialized environment – a single environment for all developers on the project to have only one test setup to manage – with automated tests that can be run in a short time frame so that you can obtain the result “PASS” or “FAIL” quickly, during the day.

Continuous integration therefore requires several concomitant elements:

– automation of all functional and technical tests of the software component, with execution that can be performed over a very short period of time, of the order of a few hours, so as not to slow down development;

– automation of software component configuration management actions in order to be able to implement rollbacks in the event that developments prove to generate defects;

– a strong reactivity of the production team in fixing identified defects.

Continuous delivery (CD) is related to continuous integration and automates the delivery of changes according to frequencies – daily, weekly, etc. – according to the needs of the organization. Delivery actions are automated.

Continuous deployment (CD) extends continuous delivery to automatically deliver developed assets to customers. There are no more delivery milestones, and developers can focus on ongoing developments.

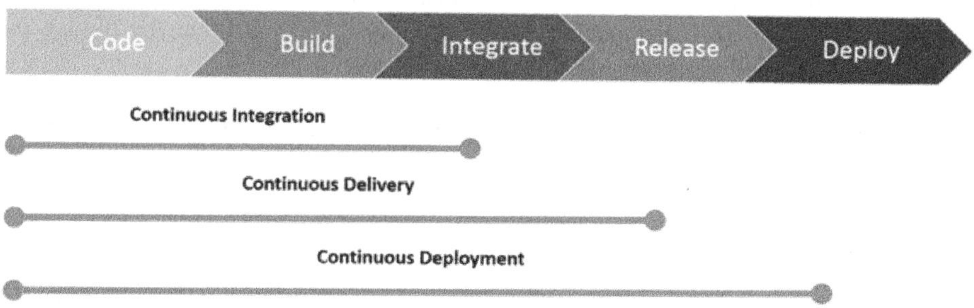

Figure 4.3. *Integration, delivery or continuous deployment*

Continuous integration (CI), continuous delivery and continuous deployment activities involve continuous and multiple evolutions of software releases. It will therefore be necessary to ensure that the components, products or systems are able to download software updates to these components, products or systems either in an automated way (e.g. during periods of inactivity of the component, product or system) or manually during maintenance periods. Similarly, it will be necessary to ensure that the loads update the version number of the component, product or system in order to guarantee their traceability in configuration.

Full automated testing coverage – functional and technical – is the only way to ensure positive continuous integration. It will therefore be necessary to design an exhaustive test coverage and maintain it throughout the project. Executions of test campaigns should provide automated reporting to decide whether or not to continue with the delivery and deployment of software.

4.5.2.6. *Automated acceptance tests*

Automated acceptance testing is an essential component of continuous delivery within DevOps strategies. However, it is a practice that is not widely implemented (less than 3% according to TestDrive-UAT) due to the complexity and the effort required to achieve a convincing result as soon as there is significant complexity.

The complexity comes mainly – in the case of systems-of-systems – from the multiplicity of components, products and systems interacting with each other. Some of these products have mechanical interfaces while others are electronic. If we really want to automate the acceptance of a physical component – or of the physical interface of the component – it is necessary that the test management system can interact "physically" with the component, to push on or to turn the buttons, etc. If we wish to validate the results of actions, we must be able to "see" and compare the result obtained with an expected result. We quickly understand that the human is more suitable – for the moment – than a suite of robots.

The main difference between understanding automated acceptance testing in a DevOps environment and automated acceptance testing in a systems-of-systems environment is that, in the DevOps environment, acceptance is software, whereas in the system-of-systems environment we include physical systems.

Some acceptance tests can be automated and are therefore good candidates to include in (non)regression campaigns. On the other hand, it is important to note that acceptance tests are designed early on the basis of the requirements and that they are carried out late, just before delivery. They will therefore undergo all the changes in requirements, business and technical, throughout the project. Any automation of these acceptance tests will depend on these changing requirements, so maintenance of these automated tests will be important.

It is also important to note that the automation of acceptance tests aims to ensure coverage of all combinations of data, activities and processes – similar to nominal and/or exceptional use – which generates immense number of combinations. Automated acceptance tests designed by test tools capable of generating the combination of valid and invalid paths are possible, but it is necessary to identify the skills and the flaws of the tool cases.

Finally, it is necessary to remember that acceptance tests are both very close to the business and of significant complexity. So, the workload for those involved in these tests is very high – due to the volume of combinations to be considered – and very complex. It would take individuals who have both a very good understanding of the business (e.g. product owners), a good ability to program test tools (e.g. developers) and a good understanding of test techniques (e.g. testers) such as data-driven testing, keyword-driven testing, orthogonal arrays, etc., including their advantages and weaknesses.

The measurement of the coverage of automated acceptance tests allows having a certain level of confidence – generally insufficient – in the level of quality of the system-of-systems. It is recommended to supplement the automated tests with exploratory manual tests performed by experienced users capable of stressing the system in rare and problematic configurations.

4.5.2.7. *Systems tests*

Regardless of the selected development model, systems testing takes place before user acceptance testing when all components, products and systems are integrated with each other. This implies that they are performed – for a given release – at the end of development and testing of that release, so very close to user acceptance testing. Any anomaly identified at this stage will therefore impact delivery to customers, either as a limitation or as a delay in delivery.

For systems-of-systems, an environment that best reproduces the real environment will be necessary. This implies coordination between the system-of-system project, on the one hand, and all the systems, flows and interfaces (e.g. ETL or EAI) and applications – legacy or not – which connect to it, on the other hand. somewhere else. In this environment, data must be in sync with each other and include historical data to replicate real life.

The measurement of system test coverage, like the measurement of automated acceptance test coverage, is difficult to achieve and focuses mainly on the coverage of the requirements defined for the system-of-systems.

4.5.2.8. *API tests*

APIs (Application Programming Interface) are interfaces defining the interactions between components, frequently between clients and servers. A component defines the various requests it can process, how they are received, the data formats to use, the conventions to follow, the results provided, etc.; other components use this information to interact with the component and use the services it provides. These APIs allow modular programming.

The difficulty in API testing lies in the opacity of the processes accessed by these applications (the content is only accessible via documented interfaces), and the complexity and combination of these interfaces. In some cases, data external to the components can be accessed (e.g. public or shared data) without the tester being notified, which can generate side effects and regressions.

APIs have no user interface, so testing must be done in isolation via drivers, mock objects or a test framework that can (1) send information to those APIs, (2) retrieve the information provided by the APIs and (3) compare the obtained responses to the expected responses. Testing APIs at the unit test level is easier because APIs and their interfaces rarely change. These tests are often used in DevOps and when setting up CI/CD (Continuous Integration and Continuous Deployment).

Testing of APIs is based on the published description of their behavior. In the framework of systems-of-systems, it is possible to test APIs through user interfaces (front-end) and client systems of these APIs.

4.5.2.9. *Tests and containers*

Containers are sets of software components, libraries and configuration files, isolated from each other and communicating with each other via well-defined channels. Containers are run by a system runtime kernel and therefore use fewer resources than virtual machines. The concept was developed in 2013 by Docker Inc.

The difficulty in testing containers lies in the opacity of these applications (the content is only accessible via documented interfaces), the fact that some of these containers are developed by other organizations, and the complexity and combination of these interfaces. In some cases, data external to the components can be accessed (e.g. public or shared data) without the tester being notified, which can generate side effects and regressions. The use of Docker and containers is increasingly widespread in the Linux world.

Containers have – generally – no user interface, so testing must be done in isolation via drivers, mock objects or test frameworks to be developed separately to (1) send information to these containers, (2) retrieve the information returned by the containers and (3) compare the obtained responses to the expected responses. Testing containers, at the level of unit tests and component integration tests, is more reasonable because the interfaces of containers rarely change. These tests are often used in DevOps and when setting up CI/CD (Continuous Integration and Continuous Deployment).

4.5.3. *Hypothesis*

4.5.3.1. *eXtreme Programming (XP)*

Using XP involves actually implementing the various XP practices. For example:

– there actually is a customer representative within the team and they are able to respond to requests from team members;

– the team uses pair programming and automates unit tests in order to filter anomalies as early as possible, including during the integration phases;

– testers are involved in the team and actively participate during the planning, requirement analysis and test design phases;

– acceptance and regression tests are performed on all modules that may be impacted by the code designed or by the data used by this code.

4.5.3.2. *Scrum and Nexus*

According to Scrum.org, Nexus is based on and extends Scrum. Nexus allows teams to collaborate to deliver a complex cross-platform product in short and frequent cycles without sacrificing quality and adding complexity, without deviating from the basic principles of Scrum. Nexus presupposes experience and practice of Scrum. The basic elements are:

– identify the teams that will participate in the Nexus;

– form an initial Nexus integration team;

– implement a unique product backlog;

– define a unique DOD (Definition of Done);

– identify a sprint cadence for the Nexus.

As with Scrum, the Nexus method alone is not enough to guarantee the success of the project, but it does identify and remove the dependencies created by the complexity of the system-of-systems. Scrum practices will need to be reviewed to adapt to the increased complexity of a Nexus. The main assumptions are therefore a correct, consistent and complete implementation of Nexus and Scrum within the organization.

4.5.3.3. *SAFe*

One of the major elements of SAFe is the synchronization of trains (ART for Agile Release Trains) or projects in progress. If one or more projects do not meet the iteration durations, the result will be waiting times before the possibility of going into production or moving to the next stage. It is important to ensure – from PI planning onwards – that activities are correctly assessed and prioritized.

4.5.3.4. *KANBAN*

The representation by a KANBAN table implies that the content of the various tasks in each column of the table is clearly and exhaustively defined. This can be done by a detailed definition of DONE as well as by a definition of READY (DOR,

Definition of Ready), that is, the minimum level of quality of the components, products or systems at the input of a column of the KANBAN.

Among the other assumptions related to the use of KANBAN boards, we have:

– the teams are stable on all iterations which allows us to compare the velocity;

– processes are complete, understood and followed;

– the environments are stable and the data is mastered.

In addition, it is necessary to note that a KANBAN board is only a tool for managing sequences of activities, identifying possible bottlenecks and monitoring the progress of work packages. The KANBAN board will not help you define or predict the completion time of a project, task or sprint.

Another important element to remember with regard to the use of a KANBAN board is that the board only describes the tasks to be carried out and does not provide information on the duration of these tasks, therefore on the capacity to complete the tasks in one iteration. In order to avoid the generation of bottlenecks, it is recommended to define, for all tasks or activities, a maximum number of WIP (Work In Progress) which becomes a limit to the number of KANBAN tickets that can be processed simultaneously.

A KANBAN board makes it possible to note the duration of each of the tasks and therefore to have a good idea of the time taken by a ticket to cross the board. This gives velocity information that can be used in the future.

4.5.4. *Calculation methods*

4.5.4.1. *eXtreme programming*

eXtreme programming assumes that unit and integration tests are estimated by developers in the design load of user stories, use cases or features. On the other hand, it is necessary to add – separately – the test loads for the system tests and the acceptance tests. In their book (Crispin and House 2003) Lisa Crispin suggests counting between 20% and 50% of the development time with an average of approximately 33% of the development time. The time required for analysis and coding is included in this development time. If the project is new or uses unfamiliar technologies, then consider higher system testing or acceptance loads.

4.5.4.2. *Scrum and Nexus*

Scrum – and Nexus – being specific to each development team, the definition of charts applicable to several teams is difficult.

The most commonly used measures are:

– number of user stories delivered per sprint;

– velocity of the team to treat the user stories.

Other measurements that can be used are described in Chapter 6. It may be easier to measure the velocity specific to your project over one or two sprints, then use that, even if it means adapting it if it turns out that the information is not correct.

4.5.4.3. *SAFe*

Implementation of SAFe can be carried out at several levels:

– an implementation of the essential elements, with three main skills: technical agility and agility of the team, an agile delivery of the product and a lean and agile direction;

– an implementation for broad solutions, including – in addition to the essentials above – a solution delivery competency for the enterprise with specific roles, artifacts and events;

– a portfolio implementation that includes, in addition to the essentials, lean portfolio management (including funding, operation and governance), a culture of continuous learning and organizational agility, and roles, events and artifacts;

– the most complete implementation of SAFe at the organization level, including both essential elements, broad solutions and portfolios.

The measurement will be determined by the percentage of correctly implemented elements, artifacts and roles.

4.5.4.4. *KANBAN*

The usual measure with KANBAN is velocity. That is, the speed at which the card moves from leaving the BACKLOG column and through each of the columns of the KANBAN board to the DONE column. It is obvious that we will take the average of the speeds, with identification of the minimums and maximums. We realize that the velocity can only be compared within a single KANBAN board, but not between boards because the number of columns and the tasks defined for each column will be different.

Velocity also depends on the number of resources assigned to perform the tasks on the board: any reduction in the number of resources assigned or any change in the resources – and their skills – will have a negative effect on the velocity.

4.5.5. *Synthesis*

Advantages	Disadvantages
Agile methods allow rapid application developments, if all the principles are correctly understood and implemented.	The optimizations proposed by Agile methods tend to focus on local optimizations, without identifying the dependencies and impacts that result from an in-depth study of the consequences of each evolution.
Adaptation of developments according to the level of maturity of the design documentation. It is not necessary to have finished writing all the design documents to start developing. This implies an upstream analysis of all the dependencies and a design in a consistent order with the prerequisites. Otherwise, developments may expect each other according to their dependencies.	The definition of the content of the sprints must be consistent both in terms of the team's ability to achieve and in terms of the design of complete functionalities. The main risk is to postpone heavy and/or complex developments until later. This increases the risk of delays and increased costs.
The practices to be implemented vary according to the risks and the choices of the team.	Any delays and changes in development time will impact the ability to deliver the feature within the expected time frame, with likely side effects on subsequent iterations and higher levels of testing.
	Increase in technical debt if no refactoring.

4.6. Selecting a multi-level methodology

In system-of-systems, it is rare for all the partners to use the same mode of development and/or completely synchronized sprints. Several test methodologies will therefore have to be used to cover all the risks and all the challenges associated with testing the system-of-systems. You will probably have to adapt or modify the methodologies and techniques to deal with your technical, organizational and human environments, as well as the various development methodologies used.

The seven basic principles of testing remain applicable in system-of-systems testing: test closest to where defects are introduced, realize that it is impossible to test all combinations of conditions, paths, and data, etc. If you combine several methodologies, you must properly understand the advantages and disadvantages associated with each of the methodologies that you – and your subcontractors – implement.

The risk-based methodology (RBT) is recommended and can be combined with the requirement-based methodology (TBX), especially if the RPN calculation method is used for requirement prioritization. Including review and inspection activities of design documents (requirements, general and/or detailed specifications, codes, test cases, etc.) and continuous improvement actions (based on metrics) is just common sense and should be generalized. In the coding phases, code analysis by static analysis tools and the use of TDD make it possible to detect many defects before the dynamic test phases, thus reducing the duration of the tests. These early testing activities should also be generalized.

4.6.1. *Hypothesis*

In our choice of test methodologies to implement, we will have to take into account the ability of the solution to adapt – automatically – to the unforeseen elements that we will encounter:

– the "product" risks, their criticality and their impact as well as the means of detecting and limiting them;

– the "project" risks and how to anticipate or mitigate them;

– anything related to the management of requirements, their follow-up and their conversion into codes and tests to validate nominal operations and the absence of abnormal operations;

– anything that may relate to the volume of defects, their late or early detection, and the impact of their corrections on both the ability of development teams to deal with them, as well as the ability of regression tests to check and/or validate them;

– anything that can lead to loss of time, rework or unnecessary back and forth (e.g. anomalies or defects identified late), so as – as far as possible – to respect "lean" principles and the recommendations of Deming (2000) and Goldratt and Cox (2001);

– if we plan to use testing tools, what confidence can we put in their use, in their ability to detect false positives and avoid false negatives?

4.6.2. *Calculation*

Given the certainty that the methodologies chosen will be a combination of several methodologies, it is not possible to define a reliable calculation method, but we can consider providing margins of error to the calculation methods of the main methodologies selected.

To ensure the most valid calculation possible, it will be necessary, among other things, to consider:

– the typology of the software and the system-of-systems envisaged (e.g. Web, commercial, military, etc.) which will impact both the complexity and the level of documentation;

– the size of the system to be developed, and – for the software – its size in function points corrected according to the complexity;

– the complexity of the organization of the system-of-systems, its co-contractors, their development methods (traditional, Agile, Scrum, SAFe, ITIL, Prince2, CMMI, etc.).

The recommended solution is to compare the calculated values obtained with orders of magnitude such as those provided in the works of Capers Jones.

Management of business risks, direct or indirect, is the major element and the main constraint. This involves the identification of all risks and all requirements (which are risks if not covered), including regulatory or normative requirements.

Budgetary and calendar constraints impact the level of coverage of risks and requirements by tests. This way of doing things can lead to the decision not to cover certain risks (or requirements) considered minor.

The volume of defects and the complexity of the applications make it possible to process the corrections in a reasonable time and to carry out one or more phases of regression and acceptance tests before going into production. This implies that the development of new features must stop well before the system testing phase.

The level of quality of the defect reports is sufficient, and the rejection rate of defect reports is low.

The frequency of use of the feature by users and the impact of the defect – specifically its frequency of occurrence – for users and the presence or absence of easy-to-implement workarounds.

4.7. From design to delivery

Often, build and test teams limit their efforts to delivering software to the next team, which then releases the software to users. There is therefore a clear separation between the DEV domain and the OPS domain. The DevOps concept links these two areas by considering that they form a single value chain. To do this, DevOps offers the most complete automation possible of verification and validation activities (static analyses and tests), as well as of version and environment management activities.

To ensure that a reliable flow is created from design to production (from Dev to Ops), it is necessary to use environments that are closest and most representative of production. Given their number and complexity, it is recommended that these environments be created automatically, ideally on demand from scripts or configuration information managed in configuration management, without human intervention from operations.

In general, this implies that any changes made in production must be carried over to the development and test environments.

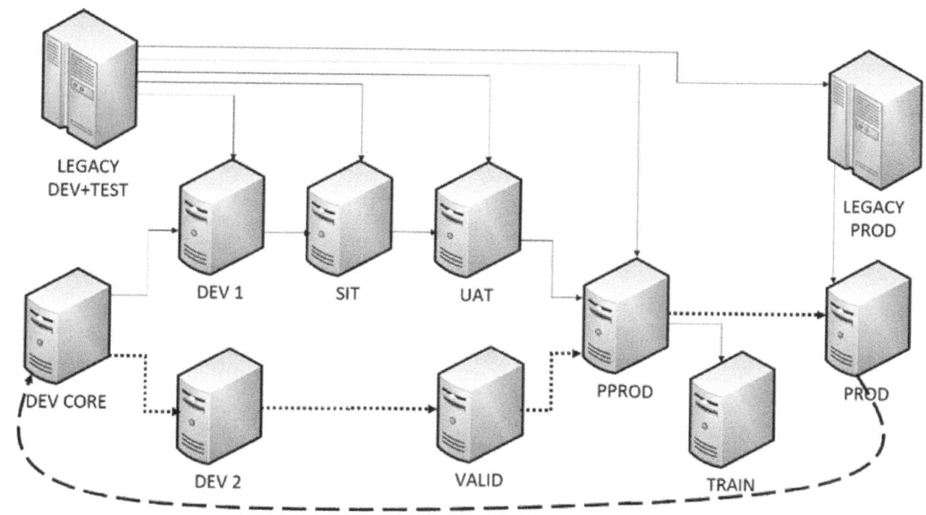

Figure 4.4. *Design – delivery and feedback*

In Figure 4.4, the planned evolutions of the applications of the new system go from DEV CORE to DEV1 and then SIT, UAT and PPROD before arriving in PROD. The urgent updates (patches and correctives) of this new system go from DEV CORE to DEV2 and then VALID and PPROD before arriving in PROD (dotted lines). This second path does not perform the regression and integration tests performed in the branch comprising the planned upgrades. It is therefore possible that certain developments or corrections are not correctly transferred from the production environment (PROD) to the other environments. It is therefore necessary to ensure that the feedback from PROD to DEV CORE is implemented (dashed curve), then goes through DEV1 and then SIT, UAT and PPROD.

For the software to provide users with the expected added value, it must be ensured that the delivery works properly. The target delivery environments should therefore be used to perform application acceptance testing. Continuous delivery systems and automated test systems should be considered in this case.

5

Quality Characteristics

The quality of a component, product or system and of a system-of-systems is not limited to its ability to do what is required of it – the functional aspect – but extends far beyond: it must ensure the security of data, the speed of execution, the volume of resources required, the ease of maintenance of its components, products or systems, etc. Quality is also perceived differently depending on the positioning of each (designer, supplier, user, etc.). Not to forget anything, there are various standards that define quality characteristics applicable to systems with preponderant software and systems-of-systems including software. Some of these standards are international and generic such as ISO25010, 2011, and others are specific to specific fields such as health (FDA Title 21 CFR Part 820), aeronautics (DO-178C/ED12C) or space (e.g. ECSS standards).

The test manager must ensure, in addition to compliance with the functional requirements, that the applicable standards are followed. Non-compliance with non-functional quality characteristics is almost always a source of dissatisfaction for users, while non-compliance with standards can lead to a ban on marketing (e.g. in the medical or aeronautical sectors). If a calculator gives the correct result (Feature) but the calculation time is too long (Performance), the calculator will be considered unusable.

Ensuring that each quality characteristic and its relative importance is identified involves an analysis – by business analysts or business specialists – of the applicability of the quality characteristic being considered. As previously defined, the test manager must identify the applicable standards for the system-of-systems in question: space systems are not subject to the same constraints as railway systems; therefore, the applicable standards – and the quality characteristics to be considered – will be different.

The quality characteristics to be processed must be defined from the start of the project or at the latest from the start of construction. Some business areas (e.g. aeronautics, space, rail, medical, automotive, etc.) will involve the coverage of many quality characteristics specific to these businesses, in addition to the generic quality characteristics described below.

5.1. Product quality characteristics

The product quality characteristics listed below come from the ISO25010 standard and relate to the quality characteristics of software:

– functional relevance, which includes: functional completeness, functional accuracy, functional relevance:

- functional completeness: the degree to which the system or software covers all the functionalities specified in terms of tasks and user objectives (therefore perhaps more extensive than the simple description of the functionalities),

- functional accuracy: the level at which the product or system provides the correct results with the expected level of precision (e.g. number of decimal places in calculations vs. precision in displays),

- functional relevance sub-characteristic of functional relevance of the ISO25010 standard: the degree to which the system, product or software facilitates the accomplishment with the specified tasks and objectives (e.g. functionalities presented to the user are limited to the needs, without including unnecessary features);

– performance and yield, which includes: time behavior, resource utilization, capacity:

- *time behavior*: the degree to which technical (e.g. response time, latency, etc.) or functional behavior varies over time,

- *use of resources*: evaluation of the use by the system or the software of the volume of resources used (memory, CPU, bandwidth, disk space, etc.),

- *capacity*: the degree to which the maximum limits of the product or system parameters reach the requirements. These parameters can include the elements to be backed up (volume of data), the number of concurrent users, the bandwidth, the size of the databases or the capacity to process transactions;

– compatibility, including: coexistence, interoperability:

- *coexistence*: the degree to which the system or software can coexist with other applications and/or systems, and identification of incompatibilities with other applications or systems,

- *interoperability*: the degree to which components, products or systems can exchange information and use the information exchanged. This interacts with the behavior over time because the data must be used in good time;

– usability, composed of: recognition of relevance, learnability, operability, protection from user errors, aesthetics of the user interface, accessibility:

- *recognition of relevance*: the degree to which users can recognize that the system, product or component meets their needs. This may include the presence of documentation (whether online or not), tutorials or other informational materials,

- *learnability*: the degree to which the product, system or component can be used by users to achieve specified learning objectives, in order to use the product effectively, efficiently, without risks and providing satisfaction in the context it is used,

- *operability*: the degree to which a product, system or component has attributes that make it easy to operate and control,

- *protection from user errors*: the degree to which the product, system or component protects users from the introduction of errors,

- *aesthetics of the user interface*: the degree to which a user interface is pleasant and satisfies the user,

- *accessibility*: the degree to which a product, system or component can be used by people with the greatest number of characteristics and capacities to achieve a specified objective in a context of use;

– reliability (ability to perform specified functionality under specified conditions for specified duration) which includes: maturity, availability, fault tolerance, retrievability:

- *maturity*: the degree to which a product, system or component achieves reliability requirements based on nominal operation,

- *availability*: the degree to which a product, system or component is operational and available when it needs to be used. Externally, availability can be assessed by the proportion of time the system, component or product is active. Availability is therefore a combination of maturity (which takes into account the frequency of failures), fault tolerance and recoverability (which deal with the duration of unavailability following each failure),

- *fault tolerance*: the degree to which a product, system or component performs as expected despite the presence of hardware or software faults,

- *recoverability*: the degree to which, in the event of an interruption or failure, a product or system can recover the directly affected data and restore the expected state of the system (or product). Following a failure, a system may remain unavailable for a significant period of time, which is determined by recoverability;

– security composed of: confidentiality, integrity, non-repudiation, responsibility, authenticity:

- *confidentiality*: the degree to which a product or system ensures that data is accessible only to those who are authorized to have access to this data,

- *integrity*: the degree to which a product, system or component prevents unauthorized access to, or modification of, computer programs or data,

- *non-repudiation*: the degree to which actions or events can be shown to have taken place, so that these events cannot be repudiated in the future,

- *accountability*: the degree to which the actions of an entity can be traced uniquely to that entity,

- *authenticity*: the degree to which the identity of a subject or resource can be proven to be that intended;

– maintainability (including aspects of correction, improvement or adaptation, including aspects of installing updates), which includes: modularity, reusability, analyzability, modifiability, testability:

- *modularity*: the degree to which a software product or system is composed of components such that modification of one component has a minimal impact on other components,

- *reusability*: the degree to which a component, product or system can be used in more than one system or to build various other products or systems,

- *analyzability*: the degree of effectiveness and efficiency with which it is possible to assess the impact on a component, product or system of a planned modification to one or more of its components, or to diagnose the shortcomings or causes of failures as well as the parts to be modified. The implementation may include self-diagnostic mechanisms (BIT, CBIT),

- *modifiability*: the degree to which a component, product or system can be modified effectively and efficiently without introducing defects or degrading the quality of the existing product,

- *testability*: the degree of effectiveness and efficiency with which test criteria can be established for a component, product or system and tests performed to determine whether these criteria have been met;

– portability, composed of: adaptability, installability, replaceability:

- *adaptability*: the degree to which a product or system can be effectively and efficiently adapted for different hardware, software, uses or operational environments. We will include the size of displays (responsiveness), memories, transaction volumes, etc.,

- *installability*: the degree to which a product or system can be successfully installed and/or uninstalled from a specified environment,

- *replaceability*: degree to which a product can replace another software product specified for the same purpose in the same environment.

In a system-of-system, a software-intensive system or an application, ensuring that the right level of quality is achieved requires setting up – for all relevant characteristics – quantified and measurable objectives (metrics) and a measurement policy that periodically reports the measured values for analysis purposes.

In general, for a system-of-system, all quality characteristics are important. It will therefore be useful to provide tests covering all these quality characteristics, in addition to the specific functional elements of the system-of-systems.

5.2. Quality in use

The ISO25010 standard proposes the following characteristics for quality in use:

– effectiveness: accuracy and completeness with which users achieve the specified objectives;

– efficiency: resources spent in relation to the accuracy and completeness of the achievement of objectives by users;

– satisfaction, including: usefulness, confidence, pleasure and comfort;

– freedom from risk, which includes: economic risk mitigation, health and safety risks mitigation, environmental risks mitigation;

– context coverage, which includes: context completeness and flexibility.

Ensuring that the system-of-system, the preponderant software system or the application, provides the level of quality in use requires the establishment of quantified and measurable objectives (metrics and SLAs) and a policy of taking periodic measurements and analyzing these measurements in order to identify trends.

5.3. Quality for acquirers

Quality for those who acquire software systems, products or services from a supplier includes:

– end-user satisfaction, that is, the ability of the system (or system-of-systems and its components) to provide the expected service within the expected time;

– compliance with control and regulatory bodies, by providing evidence of compliance with applicable standards and regulations;

– the cost, both the cost of development and installation, and also the cost of use, maintenance, upgrades, training, platforms, etc., that is, the TCO (Total Cost of Ownership);

– compliance with contracts and deadlines.

5.4. Quality for suppliers

The quality seen from the side of the suppliers, individuals or organizations, who contract with a purchaser to provide a system, software product or software service, can be estimated based on:

– functional compliance;

– reliability and compliance with reliability standards;

– usability compliance, including compliance with standards for people with disabilities;

– efficiency compliance with applicable standards;

– number of cycles for a modification;

– maintainability compliance;

– portability compliance.

In addition to these criteria mentioned in the standard (ISO9126-2 2003) other criteria must also be considered:

– user functional satisfaction criteria;

– the ability to install and improve systems, products, components or software;

– the control of versions and evolutions of versions and their deployment on the system-of-systems.

5.5. Quality for users

We saw in section 5.2 that the standard (ISO25010 2011) proposes three levels of user: primary users (who interact with the system to achieve the primary goals); secondary users (who provide support to the content provider, the system manager/administrator, the security manager, the maintainer, the analyser, the porter and the installer); and indirect users (who receive the output but do not interact with the system). A detailed table is available, providing examples of user needs for quality in use.

We therefore understand that the quality for users, whether direct or indirect users, may vary from what is envisaged by production teams. This is especially true when a component, product or system is used for something for which it was not originally intended.

In the context of systems-of-systems, several other characteristics are also important:

– the nominal availability, that is, when all is well, of the system-of-systems;

– availability in degraded mode, that is, when one or more components are damaged or unavailable. The objective here is to ensure that the service provided by the system-of-systems remains available, effective and efficient even if some of the nodes that compose it are faulty or absent (e.g. military systems in the event of conflicts);

– interchangeability, scalability and maintainability with reduced staff. We will seek here to replace hardware components – and software – without the need for specialized personnel, in order to guarantee or improve the functioning of the system-of-systems (e.g. plug-and-play type replacement without stopping the system, automatic detection of the availability of a node, etc.);

– the interoperability of the system-of-systems with other systems or similar systems-of-systems implemented by allies or third parties, without losing security.

These specific system-of-systems characteristics can be adapted and augmented depending on the environment in which the system-of-systems will operate.

5.6. Impact of quality on criticality and priority

Quality often depends on our point of view. For example, the non-display of a calendar can be problematic, and if it is used to schedule sales meetings or maintenance appointments, the impact will be different: a sales meeting can be for the signing of a contract or for a simple visit; a non-respected regulatory maintenance appointment may involve significant penalties or even a flight ban (the case of an airline).

Quality characteristics are all important in one way or the other. It is important that the assessments of quality and criticality by the people who estimate the severity of the defects and the importance of the characteristics are not called into question by the co-contractors, even if it is often in their interest to respect SLAs or delivery deadlines.

5.7. Quality characteristics demonstration

It can be easy to imagine how to test functional quality characteristics: the functionalities are described in the requirements, tests are designed to exercise each of these functionalities, and these tests are executed at the latest during system tests and product tests, acceptance of each component, product or system comprising the system-of-systems. It is clear where to position – in the project tasks – the functional test activities. However, how do we test the other quality characteristics – non-functional – such as those of the standard (ISO25010 2011)? How do we correctly test the "behavior over time", what should this behavior be and when should it be tested?

5.7.1. *Two schools*

Two schools of thought clash:

– One proposes to start with the design of the functionality and then to focus on the non-functional characteristics. This ensures that the functional concept is good before optimizing the technical characteristics. On the contrary, this solution can hardly improve the aspects of performance if they are dependent on the functional logic.

– The other proposes to define both functional and non-functional characteristics, in order to test the non-functional characteristics of components and parts before testing the functional aspects. This solution makes it possible to ensure the optimization of the components that will make up the functionalities, but certain non-functional aspects will be tested before the functional aspects, which may seem absurd.

Each school has its pros and cons. The disadvantages of the first are that the proposed functional solution can lead to decisions that are difficult to question later. The disadvantages of the second are that it is not directly seen whether the solution can give the expected functional result. Let us take an example to illustrate this, with a series of requirements (and quality characteristics) and choices that can be considered from these requirements.

We must analyse the thickness of glass bottles coming out of a mold. So we think of various solutions to measure this thickness, with either mechanical or optical solutions, in order to identify whether the thickness meets the criteria depending on whether a green or red light is displayed. The mold comes out of many bottles at a time, and we could consider testing only one sample, but how do we choose it? It was only then that we were informed that the bottles all had to be tested and that over 600 bottles per minute came out. We realize that this last information completely changes the way of managing the subject: exit the mechanical solutions, exit the red/green display, the flow rate of the bottles – 10 per second – requires fully automated solutions.

We therefore see that the quality characteristics "Performance", "Operability" and "Maintainability" become very important constraints, and that it will be necessary to test them exhaustively. In this specific case, the second school of reflection will probably be better suited than the second.

5.7.2. *IADT proofs*

Proving the conformity of a system or a subsystem to requirements is not always achievable by testing. How to prove by tests that a software was written in a certain language? In fact, we must make sure – through static tests – that the language used is the one requested.

For systems-of-systems, some organizations define the types of evidence based on the acronym IADT for Inspection, Analysis, Demonstration and Testing. These terms do not have the same meaning here as they do in the world of simple software testing:

– Inspection here is a check of the appropriate documentation and/or equipment, product, subsystem or system, to verify that the item conforms to the requirement and design documentations. It is a control by a human using only documentation, without obligation of instrumentation.

– Analysis is a verification of the specific requirements by technical evaluation and representative data that can be derived from analysis, calculation and/or models. It is a theoretical control where the expected result is obtained by extrapolation and/or analysis of a sample.

– Demonstration is the verification by observation of a sequence of operations of a specified requirement. This verification is not instrumented, but the results must be consistent with the action and its context. This context is not defined by values, and used only to obtain a consistent result. Functional testing in a test environment is part of this.

– Test here is the verification of a specified requirement by appropriate measures under conditions defined in accordance with test procedures. The context is defined with values and the test must provide only one result.

These types of evidence can be used when designing test strategies and test plans to determine the best time – in the system-of-systems design cycle – to provide the necessary evidence. In the case of software and software-intensive equipment, Inspection and Analysis may be determined by static testing (architecture reviews, system analysis, tooled code reviews or analysis), while Demonstration and Testing will require evidence provided by functional and non-functional tests on one or more representative models (Demonstration) or on a component (e.g. prototype or first component produced), in the most representative environment possible (Test). It should be noted that the provision of evidence can be carried out in an environment representing only part of the target environment: testing the behavior in weightlessness on Earth is not possible; therefore, the demonstration and test activities will have to limit the effects of gravity if it is not possible to completely eliminate it (e.g. testing the space component in a swimming pool).

5.7.3. *Other thoughts*

"Prevention is better than cure" is a saying that also applies to monitoring quality characteristics. On the one hand, it is more profitable to avoid introducing faults than to introduce faults and correct them. It is also important to integrate quality characteristics from the design phase, rather than adding them afterwards. For example, the "security" attribute involves aspects of message encryption, security of key storage (e.g. public keys and private keys), minimum level of password complexity, guarantee of message integrity (e.g. possibly with blockchain?), non-repudiation, accountability and authenticity, survivability (ability of the system to continue its mission by providing essential services, even in the presence of attacks), immunity (resistance to attacks), etc. These attributes must be integrated from the design of components, products and systems and not added afterwards. Verifications by reviews or inspections can complement dynamic test activities carried out by specialized testers.

It is therefore necessary to think about how to cover the quality characteristics from the initial design, that is, during the design phase of the test strategy.

6

Test Levels

Systems-of-systems testing can be considered at various levels, each one more detailed than those above. This vision can be considered as linked to sequential development modes, but corresponds well to the mode of design – and test – of systems-of-systems: a general objective is defined then refined and detailed in several systems, products or components of smaller size, for which manufacturing or design contracts are drawn up. Smaller elements can be developed according to other design patterns (e.g. Agile, SAFe, V-cycle, Spiral, etc.) at the choice of the organization, as described in Figure 6.1.

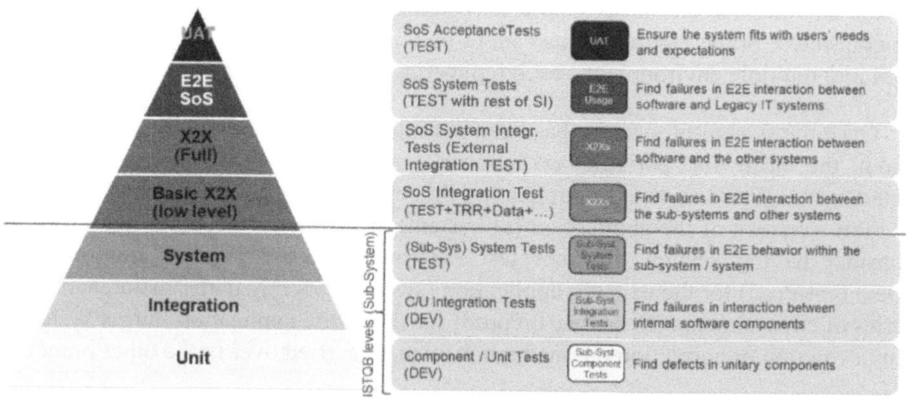

Figure 6.1. *Test levels in a system-of-systems. For a color version of this figure, see www.iste.co.uk/homès/systems1.zip*

The definition of the test levels can be based on the granularity of the objects subjected to the tests, as well as on the quality characteristics – the typologies of anomalies – which will be investigated at this level. For example, we could have software components – so the lower three levels – that are integrated into a product – so hardware–software integration tests to consider – then the product would be integrated, on an integration test bed of basic X2X type with other products to create a first-level system. The system would then be integrated – full X2X test level – with the other systems of the system-of-systems to verify the exchanges, prior to a set of system tests (E2E SoS) then – once all the tests have been successfully executed – an acceptance phase can be implemented. We see that the number of test environments will correspond to the number of test levels envisaged.

The principle is to evaluate the quality characteristics considered relevant on the component or system targeted by the tests at this level. Generally, the functional aspect is considered to be predominant, but a quality reference system (e.g. ISO25010) makes it possible to identify characteristics other than "business" functional objectives.

Like any other activity, testing needs to be effective and efficient. Effective in terms of defect detection, efficient in limiting the effort required to achieve these objectives.

This minimum number of environments is valid when designing a product or system. The production environment of a system-of-systems will be connected with all the other systems that compose it. As it is necessary to test this system-of-systems in an environment as close to reality as possible, the test environment should also include interfaces with test versions – synchronized – of the systems to be connected. Some environments cannot be completely replicated, so they will be split into multiple environments, each replicating part of the target environment.

Once the system-of-systems has been put into production, it is necessary to think about the need for periodic evolutionary maintenance and urgent corrective maintenance. This can lead to duplicating the test environment and maintaining two or more versions of the application (or system) in parallel, it becomes necessary to consider having several different test environments: a series of environments (Dev-Test-Prod) for the existing application (system currently under maintenance), a series of environments (Dev-Test-Preprod) for the future application (target system). Any evolution or correction on one branch must be carried over to the other branch.

Switching activities from one version to another – with or without data conversion – may require other additional environments.

6.1. Generic elements of a test level

All the test levels share the same set of activities and concepts, with specificities according to each level:

– a set of repositories (called "test base") defining what must be developed and tested (requirements, User Stories, etc.), as well as the expected results;

– test targets, which will undergo the tests. These targets (components, products or systems depending on the test level) will have to interface with the other applications, components, products or systems of the level (in the relevant versions);

– an environment where tests will run, receiving validated developments from lower levels, and integrating with the applications needed at this level. In addition to the presence of the environment and the components, products or systems with which the test target will interact, it is necessary to include all the configuration of these elements;

– test objectives for this level. We will mainly look for failures in the behavior of the component, product or system as well as in the messages they send or receive;

– test data necessary for the execution of the tests;

– resources, test tools and actors;

– testing processes, adapted to the level;

– test methods and techniques adapted to each level;

– specific risks, such as:

- late delivery of elements describing the expected functioning – detailed and unambiguous – of the designed components or undocumented modifications. This includes the description of the expected results of the execution of the tests,

- lack of a properly sized, parametrized, configured test environment, with authorized access for testers,

- lack of usable test data for test execution,

- missing applications with which the component, product or system must interface and exchange,

- absence of the component, product or system to be tested.

6.1.1. *Impacts on development cycles*

Development cycles have little impact on the types of tests: all should be carried out because each focus on a typology of defects. Whatever the development cycle, we should start by doing reviews (of specifications, requirements, User Stories) and static analysis (of the code), then we should perform unit testing (also called component testing) of the components, then component integration test, then system test and system integration test, before doing acceptance testing.

In an Agile development model, each sprint after the first will ensure that developments do not introduce regressions into what has been developed previously. This will involve performing (non)regression testing and sometimes automated testing. When implementing continuous delivery solutions (e.g. DevOps), it will be necessary to ensure, through systems integration testing and system-of-systems testing, that end-to-end functionalities continue to work and are not impacted. Performance, security and usability testing will also be periodically required.

In an Agile development model, the various levels of testing are performed sequentially for a requirement (or user story or use case or feature), within a sprint.

In a sequential development model, each level of testing is separate, focused on finding defects of different types.

6.1.2. *Methods and techniques*

In the context of system-of-systems testing, all the usual test techniques can be used. For lower levels (e.g. component testing and unit testing) structural techniques (white-box testing) can detect anomalies that would not be detected by functional testing (black-box testing). It is recommended to run the functional tests first, then measure the achieved structural coverage rate, and increase this coverage rate by designing structural tests.

Depending on the level of criticality of the software, as well as the applicable standards and regulations, the level of structural coverage to be achieved will be different. For example, we can cite the SIL (Software Integrity Level) present in the IEEE829:2008 standard and the criticality categories of aeronautical software proposed by the DO178B/C ED12B/C standard. Similarly, testing techniques and methods, as well as the level of effort associated with each requirement, will vary depending on the level of risk or criticality of the component under test. The principle is to have a test coverage that increases according to the level of risk.

For higher levels of testing (e.g. system, system integration and acceptance testing), the combination of tests and systems (and their states) is such that it is impossible to test all combinations. You will thus be able to select techniques that limit or control the combination, considering the level of risk.

6.1.3. *Fundamental principles*

Some elegant concepts, also present in the concept of "Lean", have been defined by Koomen et al. (2006) and apply to any test activity:

– Central starting point: the objective is to have one or more states from which to launch most tests. This makes it possible, if, for example, it is necessary, to identify oneself before being able to use an application, to define the central starting point as being the "identified user" state. Therefore, all test suites and all test cases will start from this state and should come back to this state.

– SETUP: the objective here is to ensure that the runtime prerequisites are ready.

– RUN: is the test execution phase including the reporting of results.

– TEARDOWN: is the phase where we will make sure to return to the central starting point.

These concepts should be implemented for both manual and automated testing. They support the concepts of "Lean" by limiting the loss of time in the execution of campaigns and by facilitating the execution of tests.

We recommend PATH techniques, both for valid and exception paths, with the use of equivalence partitions (EP), limit values (BVA) and decision tables (DT) and states and transitions techniques (STT).

6.1.3.1. *SETUP phase*

The concept of SETUP applies for a test level, and also to each of the test activities of a campaign, from design to execution monitoring. The SETUP phase of a test level will include everything necessary for the level to be tested effectively:

– ensure that functional prerequisites are present and available (e.g. risks, specifications, requirements, data dictionary, data flow between systems, products, components or software, etc.);

– ensure that resources are available (e.g. tool licenses, test data, etc.) and that individuals have the right level of knowledge on both the functional aspects and the tools that will be used;

– ensure processes are defined, complete, understood and applicable, specifically for cross-team handover, issue management and release/component management;

– identify the tasks to be carried out on the level and the loads (efforts) necessary to carry them out. Identify prerequisites, constraints and dependencies in order to be able to anticipate them;

– identify the flows, needs for data anonymization, dependencies, prerequisites, organize all the environments required for the execution of the level tests to be carried out correctly (other systems, network and configuration, applications and accounts of users, flow settings, etc.);

– prepare the data (configuration, input data, expected data, checklists, etc.), create or obtain the input data, if necessary make it anonymous, determine the sources of information to identify the expected data. If the available data is partially incorrect, the objective of the SETUP phase is to make it correct and consistent;

– obtain the test tools and the means of measurement, configure them, define the KPIs to be used, how to calculate them, what they will show and the reporting templates to be used;

– analyse risks and requirements, to define the test load (effort) according to their criticality and/or dependencies. Implement appropriate testing techniques and design the necessary test conditions and cases;

– ensure that testers have the necessary access and authorizations to use the level test environments;

– detail and automate tests where needed;

– check the operation of the test environment.

The purpose of the SETUP phase is to ensure that the execution phase (RUN phase) runs smoothly and efficiently.

6.1.3.2. *RUN phase*

The RUN (execution) phase is the one during which – in a test level – the tests are executed. The aim is to work efficiently, that is, to achieve the defined objectives – find the most important faults, demonstrate correct operation, etc. – and to work efficiently, that is, not to do the same work twice and not to do unnecessary work. If the SETUP activities have been performed correctly, the RUN activities should run smoothly. Let us be realistic; there are always problems, big or small.

In the RUN phase, we will:

– execute the test cases in the defined order of priority, tracing the execution results from the requirements and risks identified during the preparation, and up to the execution results. It will be important to also ensure technical traceability to the version of the component, product or system, as well as functional traceability;

– identify anomalies and fill them in according to the defined process, using the appropriate anomaly management tool. All traceability, priority, severity and impact information will be introduced in order to follow their progress throughout the project and the sprint;

– collect execution results and measure the progress of level testing activities, in order to compare them to the forecasts – defined in the strategy (or test plan) and refined during the SETUP phase – in order to compare the achieved with the forecast and take the appropriate adaptation measures;

– to the extent possible, restore the data to a known prior state (the initial or starting state of the test), after the execution of the tests, whether they are positive or have identified failures. These restoration activities are also carried out during the TEARDOWN.

A RUN phase may have to be executed several times, for example, in the event of the presence of anomalies or new versions of components, software, products or systems. It will therefore be necessary to identify them to ensure the correct coverage of the components by the tests.

In an Agile environment, RUN activities should be carried out for each User Story, use case or feature, and therefore will be re-executed several times to ensure that there are no regressions. It is also possible to group the execution of tests at the end of the sprint to limit re-executions. This final solution is a hybrid between Agile and sequential, not always optimal.

6.1.3.3. *TEARDOWN phase*

The concept of TEARDOWN, like those of SETUP and RUN above, applies to a test level, to a test campaign, and also to a single test. The principle is to return the test environment – and the data – to the state prior to the test to be able to rerun the test if necessary.

Once the level test campaign execution is finished, it is necessary to clean the data after archiving the results. The "cleaning" of the data will make it possible to rerun the campaigns if necessary without the results obtained being influenced by previous tests; archiving of test execution results provides proof of execution.

These activities are included in the TEARDOWN phase.

6.2. Unit testing

In the context of systems-of-systems, we will differentiate unit tests from component tests which will be detailed later. Unit testing is primarily performed by development teams to check if unit (i.e. unit testable as a function, subroutine, method or class) items are functionally correct and non-functional. Some authors call these tests subroutine tests. According to Jones and Bonsignour (2012), the efficiency of subroutine testing is around 55% (i.e. 55% of defects that can be found at this level are detected and removed). This contrasts sharply with the unit tests of the PSP/TSP process proposed by Watts Humphrey, which is more formal and more efficient (65% to 95%), or the unit tests carried out within the framework of Agile projects in XP (50%) or tests standard units (35%).

Unit tests are the first tests that developed software undergoes. Note here that the TDD, BDD and ATDD type tests do not meet the test definition, and should be considered as automated requirements. TDD, BDD and ATDD are very useful, but they are not part of unit testing. In unit testing, the level of unit granularity can vary from very small (function, method or class) to very large (application or software system). For our convenience, we can think of a unit as something that can be managed separately in configuration management:

– Use: well-designed unit tests can find anomalies and fix them immediately. Since test design techniques are rarely taught to developers, the level of effectiveness of unit tests is often low. These tests are nevertheless very important in that they make it possible to identify defects or failures very early on and to acquire a level of confidence in the quality of the components. They therefore allow a better focus of other types of tests.

– Impact: unit tests are often carried out by developers as soon as the coding is finished and allow faults to be found – and corrected – quickly. When they are not (or not well) executed, defects can only be discovered in later phases of testing, which will negatively impact efficiency. It is important to have a component test environment separate from the development environment. This level focuses on the operation of programs, subroutines or functions, during their design by developers. It verifies – mainly in the developer environment – that program-specific requirements have been developed correctly.

Tests at this level focus primarily on verifying the internal behavior of each program, subroutine or function.

In an Agile development cycle, it can include TDD, ATDD and BDD. These tests are primarily white-box tests, but may include black-box tests.

– Objects: individual software elements, classes, methods, programs, subroutines, functions, APIs, etc.

– Referential: reference documents are contracts, specifications, expressions of needs, requirements, epics, features, user stories, architecture documents, as well as the results of general and detailed analyses. It is also necessary to take into account the experience feedback reference system if it exists.

– Techniques: depending on the level of granularity of the components, the test techniques can be of the white-box or black-box type.

– "White-box" tests are performed on low-level software components (the code) using various techniques (see Chapter 9) and coverage measurement (coverage of statements, decisions, conditions, MC/DC, etc.). In general, these will be checks on how the code was developed (code analysis), and they are also tests to ensure that the analysed components work as expected, and cover all the instructions, all possible decisions, all possible conditions or even all combinations of conditions and decisions. These "white-box" techniques will require an ability to understand the code for the design of the tests.

– "Black-box" tests mainly focus on the verification and functional validation of components. In the context of system-of-systems testing, it is not always easy to determine what is the responsibility of unit testing and what is not. Depending on the perspective of organizations involved in development, unit testing – sometimes called component testing – may focus on testing classes and methods, programs, subroutines and functions. Other organizations will consider unit items to be hardware features or components that include both software and hardware.

6.3. Component integration testing

Integration tests (IT or ICT) are intended to verify the integration of components with each other, via interfaces. As mentioned by Engel (2010) a system-of-systems comprising physical components or several types of interfaces can be taken into account:

– the hardware interfaces and the exchanges that occur between two physical components (e.g. exchange of fuel between a tank and the engine);

– spatial (or volumetric) interfaces, that is, their interactions in space such as the volume used by an antenna when it moves around an axis (volume that must be left free so that the antenna is not damaged);

– energy exchanges, whether heat (consider energy dissipation to avoid overheating) or electricity;

– the exchange of information. It is mainly this type that will be studied by the software tests. We will have to consider hardware/hardware interfaces (e.g. electrical cables and connectors), hardware/software interfaces (e.g. positioning of a switch impacting the behavior of software), software/software interfaces (most exchanges between software and subsystems) and interfaces between the system and humans (e.g. GUI and also UX).

We focus on the flows and messages exchanged between components previously tested individually. Here too functional and non-functional (technical) aspects should be measured.

According to Jones and Bonsignour (2012), the effectiveness of integration tests in detecting and removing defects is around 50%. Some organizations perform daily integration testing according to Lean and DevOps principles. Integration tests benefit from static analyses and formal inspections (e.g. identification of dependencies), and are negatively impacted by values greater than 20 for cyclomatic complexity.

6.3.1. *Types of interfaces to integrate*

If we abstract from the other types of interfaces mentioned above and focus on the interfaces between software, we have three major types of interfaces, based on the type of exchange between these software:

– “Real Time” interfaces which transmit data in real time between components, products or systems, for which the verification of the data exchanged is as important as the speed of exchange. These are interfaces like those used in control systems (e.g. brakes, injection, airbags, secret code validation, etc.).

– The interfaces that we will call “Timely” where data is transferred when there is a drop in bandwidth usage, or when the transfer price is lower. These are cases where a set of transactions is grouped together before being sent (e.g. statements for maintenance after a cycle, non-urgent financial transactions, etc.).

– “Batch” type interfaces where large volumes are transmitted, often at the end of the day, to update databases, perform consolidations, etc. What is critical here is the guarantee of data processing, the ability to recover in case of failure of the system receiving the data (e.g. ensuring that each data transmitted has been correctly processed and processed only once).

Each of these three types of interfaces to be integrated is to be tested differently:

– Real-time interfaces can be tested simply by running applications, focusing on the response, on response time and impact of/on bandwidth.

– Opportunity Time and Batch interfaces can be tested through the use of test files comprising valid and invalid combinations of data to verify the behavior of target components, products or systems.

6.3.2. *Integration challenges*

Integration testing involves collaboration between at least two components, products or systems, with data exchange sometimes passing through a third system.

Two challenges need to be considered. The first is the order in which the components will be delivered to you. This is to be coordinated with the production teams so that the components to be integrated reach you before starting the integration. If one of the components arrives late, you will be forced to delay the start of your integration tests or you will have to design stubs to replace the missing component(s).

The second major challenge is to ensure the compatibility of each of the components, products or systems with each other. This can be problematic when the development cycles are not synchronized with each other: an interface modification may have been applied on one side, but must also be applied on the other side so that integrations can be carried out successfully.

Among the other challenges of integration testing – regardless of the level of integration – we have the aspects of:

– performance, response time, bandwidth and throughput;

– data integrity, identification of missing or duplicate messages;

– security and non-repudiation of data, identification of incorrect messages;

– software versions management and synchronization between components;

– data mapping and its evolution.

Each aspect will need specific tests depending on the quality characteristics to be validated, their prioritizations and the identified risks.

6.3.3. *Integration models*

Various integration models exist ("big bang", from bottom to top, from top to bottom, by transaction, by functionality or neighborhood to name only the most frequent) each with its own advantages and disadvantages. Integration testing requires both technical knowledge (development and network) and functional knowledge (business). When the granularity of the components is very important (e.g. systems within an information system or a system-of-systems), we will speak of "System Integration Tests". These system integration tests are mainly functional and verify the correct connection of the systems to each other as well as the quality of the flows (mapping, performance, etc.). When testing system integration, the potential combinatorial is huge, so you will have to consider – depending on the risk levels – limiting this combinatorial.

– Use: ensuring that components are able to exchange information correctly helps to limit worries during system tests and acceptance tests. However, this type of testing is often overlooked or sloppy. This results in concerns when integrating components, especially when a software system is introduced into an information system or a system-of-systems.

– Impact: when integration tests are insufficient, system tests are often a sequence of executions followed by stops and restarts after correction, which is not optimal. To properly implement integration testing, all integrated components, products or systems must be present and properly tested (to an acceptable level of quality). This may involve a significant investment in terms of environments. This level makes it possible to integrate programs, subroutines and functions – which have previously been the subject of unit tests – into a larger software component. Software integration tests can run in a development environment or in a dedicated test environment (e.g. build verification environment).

They mainly focus on the exchanges between integrated elements, and between these elements and the rest of the software elements already integrated.

In an Agile development cycle, TDD, BDD or ATDD type tests can be used, as well as manual tests (e.g. exploratory tests). These tests – mainly functional – should be supplemented by non-functional tests (e.g. performance or security).

Test environments can be dedicated hardware environments. In the case of VHDL component developments, these will often be simulators.

– Objects: components, products and systems to be integrated (transmitters and receivers of messages), components, products or systems in charge of information exchange (e.g. EAI and ETL).

– Referential: based on the data model and the architecture diagram linking the components, products and systems of the system-of-systems, feedback repository.

– Technical: technical tests (e.g. performance and security) and black-box functional tests.

6.3.4. *Hardware–software integration tests*

A system-of-systems, or software-intensive system, runs software on hardware. These materials can be either components developed specifically (e.g. electronic boards or circuits) or off-the-shelf components (COTS). In both cases, it is important to ensure that the software, programs, subroutines or functions that will be integrated into it work properly.

In general, we will have a specific test environment to ensure the correct integration of the software on the hardware.

The tests at this level will verify the behavior of the hardware–software couple, mainly in the form of black-box tests, from the input and output interfaces of the hardware component.

Some specific aspects may need to be assessed such as:

– diffusion of electromagnetic fields which would generate information leaks;

– sensitivity of the component – and of the application – to electromagnetic fields (e.g. strong currents, radiation, etc.), to changes in temperature, pressure or hygrometry.

– Objects: we will need hardware and software components. In the case of programmable electronic components (e.g. FPGA), the encoding of the software on the electronic component then the use of the electronic component in equipment and the integration of the equipment within a test bench are mandatory steps before ensuring the correct operation of the electronic component. At each stage, it is possible that there is fault detection and therefore an obligation to start the design again.

For programmable components, if the software can be installed by the user (e.g. mission plan in an aircraft piloting system, introduction of specific maps for flying at very low altitude), integration of the software with the hardware will also have to ensure that the software has been correctly introduced (e.g. use of verification codes).

– Referential: mainly interface descriptions and description of functionalities, feedback repository.

– Technical: technical tests (performance, security, reliability) and black-box functional tests.

6.4. Component tests

In systems-of-systems, component testing differs from unit testing in that it mainly concerns equipment comprising hardware and software and for which the tests aim to ensure that the physical, technical, thermal characteristics, etc. (e.g. radiation, etc.) meet defined requirements.

Component tests will – like unit tests – verify the inner workings of each component, either directly (e.g. via probes on the component) or via input and output interfaces.

– Objects: products or equipment comprising both hardware and software. Interfaces are both physical (sockets and connectors) and logical (API, etc.).

– Referential: the design requirements of each of the components.

– Techniques: mainly black-box techniques with test design based on component specifications.

6.5. Component integration tests

Component integration tests verify and validate exchanges between previously validated components. The objective is to ensure that the messages exchanged are correctly formed, understood by the various components and transit without undergoing – or generating – interference on the communication buses.

Component integration testing should also ensure that:

– buses do not generate incorrect messages that would be interpreted by components;

– components can be identified (so that the hardware and software configuration is mastered) and if necessary replaced (including hot swapping);

– connector length does not generate side effects in the messages exchanged (when high-level accuracy is required – such as for Global Positioning Systems – the length of connector impacts network latency and thus accuracy);

– components are isolated from electrical surges (e.g. from EMP) if necessary, for example, by inserting optical components (necessary for military hardened systems);

– latency and bandwidth are enough for current and anticipated (future) needs.

Component integration testing can be done in a variety of ways:

– Big-bang integration: all components are integrated together at one time. This integration method limits the number of drivers and stubs to be created. However, the effort required to identify an integration defect is very significant.

– Top-down integration: this integration starts with the highest level element and integrates the elements of the level directly below. Absent elements are replaced by stubs which must be developed and simulate the responses of the components they replace. As soon as the components are integrated in place of the plugs, they become unnecessary. The evolution of interface specifications involves the evolution of drivers and plugs.

– Bottom-up integration: these integrations start with the integration of the lowest level elements and the elements that are directly related to them. It will be necessary to design drivers to provide data to the components to be tested and to analyse the responses provided. As soon as the components are integrated with those of a higher level, the drivers are no longer useful. The evolution of interface specifications involves the evolution of drivers and plugs.

– Integration of components two by two uses caps and drivers and makes it possible to focus on a limited number of interfaces to test. The evolution of interface specifications implies the evolution of drivers and plugs.

– Functional integration (the components are integrated according to the functionalities provided), which makes it possible to group together in a single integration all the components impacted by this functionality.

– Integration by neighborhood (components are integrated according to their physical or logical neighborhood).

- Objects: the components, networks and component interfacing systems. As far as networks are concerned, they can be wired, optical or radio. Hardware interfaces (plugs, connectors, etc.) should also be tested for electrical reliability, including resistance, reliability and durability.

- Referential: the functional and technical specifications of the architecture, description of the interfaces between each of the components and of the system-of-systems, feedback repository.

- Techniques: see integration tests and do not forget the physical interfaces between components (sockets, plugs, buses and cabling, receivers, converters, etc.).

6.6. System tests

System tests: these tests aim to find functional and/or non-functional failures in the developed system, after integration of all the components and/or systems. If the system is composed of other systems, we speak of systems-of-systems tests. Given the increasing complexity of systems and the combinatorial explosion of tests, it is difficult to guarantee complete coverage by system tests.

According to Jones and Bonsignour (2012), the efficiency of system tests to detect and then remove anomalies is around 55%. For large projects, formal system testing can take several months. Currently, the term system testing sometimes includes security testing and performance testing. Mainly using black-box testing, system testing requires configuration management and formal anomaly management.

A specific system test environment, different from the integration test, the component test and development environments is necessary, as well as a controlled process for managing versions and anomalies.

– Use: allows you to check the end-to-end operation of the system in an environment similar to the production one. It is at this level that all the interactions of the components between them will be identified. The number of possible path and data type combinations makes complete system test coverage difficult to achieve. As system tests are carried out at the end of the development chain, the time available to correct anomalies is limited.

– Impact: the systems tests are the last tests carried out before the supply of the software for acceptance by the customer. There will therefore be little time left to correct the anomalies found and to ensure that the corrections do not introduce regressions (see (non)regression tests below).

– Objects: systems, products, equipment and components, networks (wired, optical or radio), procedures and operating manuals.

– Referential: functional and technical specifications of the systems and the system-of-systems, user experience, feedback repository.

– Techniques: black-box test techniques, black-box functional tests and system integration tests.

Integration tests in a system-of-systems focus on the verification and validation of exchanges between systems. As for any level of integration, we focus on the identification of faults in the messages and the exchanges between components of the system-of-systems. The types of interfaces to consider in system integration testing have been described previously. In system testing, we will focus on the impact of systems on one another and on the interfaces. The materials for these tests may vary:

– transmitted by cable (twisted pairs, shielded cable, etc. with different speeds);

– electrical and/or electronical;

– transmitted by radio, Wi-Fi, BlueTooth, 3G, 4G, 5G, etc.;

– transmitted by light (e.g. fiber optic);

– transmitted by other means;

– etc.

It will therefore be necessary to consider (and simulate) the disturbances that each of these media can have and the impact on the operation of the interfaces between the systems. A mapping of exchanges, data flows, from their creation to their use, including their transfer between components, products or systems, will make it possible to ensure that the data is correctly transmitted without being modified.

These exchanges can be processed by EAIs converting data and transferring it between various components, products or systems.

– Objects: systems, products, equipment and components, networks (wired, optical or radio), procedures and operating manuals.

– Referential: functional and technical specifications of the systems and the system-of-systems, user experience, feedback repository.

– Techniques: see integration tests.

6.7. Acceptance tests or functional acceptance

Customer acceptance testing is common for all outsourced projects. According to Jones and Bonsignour (2012), the effectiveness of acceptance tests in detecting and correcting anomalies is around 30%. This last phase of testing is therefore one of the least effective in identifying anomalies.

Acceptance testing activities go beyond just checking functionality for users: all quality elements can – should – be checked. If the results are functionally good but arrive too late (e.g. late triggering of an air bag), the expected operation will not be present.

These activities serve to demonstrate that the developed application can be accepted in production. These are not strictly speaking tests because they are not intended to find faults. Various types of acceptance tests exist (beta tests, pilot phases, etc.). Acceptance tests can be executed in various test environments, either limited to the execution environment of the application under test, or including all the applications of the information system – or of the -systems – with which it must operate.

– Use: these activities verify the contractual requirements and acquire proof of their correct functioning, and show the users that the application can be accepted in production. Any defects identified at this stage will have a negative impact on customer satisfaction.

– Impact: these tests are often carried out by end-users who focus on their use in the context of their business. Acceptance tests are rarely exhaustive (they do not cover all paths), formal (they do not demonstrate coverage of all contractual requirements) and the identification of complex or exceptional cases is infrequent.

– Objects: system-of-systems, systems, products, equipment and components, networks (wired, optical or radio), operating procedures and manuals.

– Referential: functional and technical specifications of the systems and the system-of-systems, user experience, feedback repository.

– Techniques: black-box test techniques, regression tests, performances tests and security testing.

6.8. Particularities of specific systems

6.8.1. *Safety critical systems*

A safety-critical system is a system whose failure or malfunction could result in serious or lethal injury, or significant damage to property or the environment. There are many examples of safety-critical application areas such as medical devices, aircraft flight controls, weapons and nuclear systems (I like Rex Black's definition of "any metal box containing humans, such as cars, planes, elevators, etc."). Modern information systems are becoming critical to security in a general sense, as financial loss and even loss of life can result from their failure. Future safety-critical systems will be more common and more powerful, and also more complex and numerous.

The very visible problems that have arisen in the field of information systems security suggest that security is also a major issue, as are the aspects of responsibility of the implementation teams.

We will therefore be required to provide evidence that all failure modes have been considered and covered. This will involve the use of recognized standards applicable to the business functional area.

As the failure of a safety-critical system can have significant impacts on property and people, it is recommended to carry out, before the acceptance tests, a detailed review – inspection – of the tests carried out at the lower test levels.

Running tests in a production environment is NOT recommended. The objective of the test being to identify defects, testing in a production environment may result in unexpected behavior of the system, with critical or lethal consequences.

6.8.2. *Airborne systems*

Aeronautical systems are safety-critical systems when they impact the ability of the crew (pilot and co-pilot) to perform their tasks. The applicable standards are: DO178B/C and the regulations put in place by the EASA.

Other standards may apply depending on the type of aircraft – airplane, helicopter, drone – used and whether these systems are used by the military or by civilians.

6.8.3. *Confidentiality and data security*

The data used in systems-of-systems is often of paramount importance to the organization: whether it is customer information (e.g. emails, telephone numbers or addresses), strategy or business performance of the company (e.g. orders, turnover, etc.) or other information to be secured.

7

Test Documentation

Let us define what we mean by "test documentation" to understand the scope of the topic. The test documentation consists of all the documents and deliverables used or generated by the test activities: the contracts and requirements, the specifications and the code are deliverables at the start of the test, while the test plans, test files, test conditions and cases, test data, expected results and results obtained during test execution are deliverables resulting from test activities. The test documentation integrates with the quality documentation (e.g. Quality Assurance Plan) of the system-of-systems and develops the required evidence to demonstrate each of the requirements (e.g. Conformity Construction Plan). We differentiate documentation from reporting, which is detailed elsewhere.

The input information for the tests and used to design the tests is a test repository, better known as the "test base". In a system-of-systems, or a software-intensive system, changes to the repository can often only take place through a process of amendments. We will speak in this case of "frozen test base".

Certain test documents, more specifically the test policy, the test strategy and the test plans will aim to define – for the company and the customers – the reasons for which we test and to describe the approach implemented, as well as work to test the system, that is, how compliance with the requirements will be implemented and demonstrated (e.g. via a Compliance Construction Plan).

In Agile development cycles, extensive documentation is often considered less useful than working software. If this vision is applicable in short developments, software products and systems-of-systems are used for many years, so a person – developer, tester, etc. – having participated in the design of these systems will never remain in the same position or on the same project throughout its existence. It will then be necessary to maintain and provide information – documentation – so that the project can be supported in the best conditions over the years. Creating documentation is seen as less rewarding than creating the application, but documentation is necessary for optimal use of applications.

7.1. Objectives for documentation

Documentation does not directly add value to the system, but it is absolutely necessary for the design of a system-of-systems – in order to clarify what must be designed and tested – and for the qualification of the system-of-systems, to ensure that it correctly meets requirements and applicable norms or standards. We can split the documentation according to its type and usefulness:

– Contractual documentation, formal or not, which includes the high-level requirements as well as all the derived requirements accepted by both parties: client and design team. In this level of documentation, at the non-formal level, I would include risk analyses and prioritization of requirements.

– The added value comes from the limitation of back and forth between the production team and the customers, and the creation of evidence that can be used in the event of disputes.

– Design documentation, including everything needed to design the application, from the definition of requirements, and from the high-level architecture to the lowest level of code documentation. This documentation is intended to describe how the software should be (or was) developed. The added value comes from saving time during the design and limiting back and forth within the production team. To this design documentation, we must add the documents that will demonstrate how the requirements will be verified (see IADT), and therefore, the type of evidence that will have to be provided.

– The test documentation includes everything that should make it possible to ensure that all the requirements defined in the contractual documentation, formal or not, have been covered. The added value here comes from saving time for the correction of anomalies and for the execution of tests, from the provision of proof that the initial requirements have been correctly developed and that contractual, normative or regulatory requirements have been covered in a way adequate. The test documentation should be sufficiently complete and exhaustive to:

- train new testers in the particularities of the application or system to be tested,

- serve as proof or evidence – a posteriori – for contractual or regulatory needs,

- allow better use, for all types of users (including administrators), of the application and serve as a basis for training.

7.2. Conformity construction plan (CCP)

In a system-of-systems, the objective of the CCP is to describe how the compliance of equipment, subsystems and systems with the higher-level requirements will be demonstrated. This compliance can be based on the acronym IADT, and describe the means implemented to demonstrate compliance. As part of this book we will focus on software compliance, through static testing activities (e.g. reviews and inspections) and dynamic testing (functional or non-functional) throughout the project.

Any high-level requirement can be broken down into one or more lower level requirements, and these requirements are subject to evidence generated during the various levels of testing.

7.3. Articulation of the test documentation

As mentioned in TMAP, customer objectives and project success criteria must be identified, so that test levels and detail of test effort are assessed and defined. The test documentation should make it possible to identify the activities to be implemented and to evaluate them.

In general, we will have:

– a master test plan (MTP), describing the general organization of the tests;

– a Lavel Test Plan (LTP), per system, subsystem or equipment, describing for the level, the system or subsystem or the equipment, how the tests will be organized, how the test activities will be synchronized between the levels and between the system under test and the other systems with which there are interactions, and how will be reported – or even synthesized – the information coming from the lower levels;

– test files identifying the quality characteristics to be covered by tests, even the risks to be covered and the way in which the risks will be covered according to their level of criticality, priority or impact;

– test documentation including test records, test procedures or scenarios, test cases, test scripts, test logs, test results or test reports, defect lists, test data, etc. These test documents can be defined for each of the test levels, and for each of the systems, subsystems and/or equipment.

Test documentation must be able to reference design and development documentation (e.g. requirements or risks). Referencing – also called traceability – aims to:

– provide evidence that the requirements (or User Stories in the Agile world) have been properly verified and validated;

– check, in case of defect identification, if the defect is linked to one (or more) critical requirement(s) or not;

– monitor the evolution of requirements and their developments or their translations by code, as well as the versions and iterations in which the requirements are carried out and made available to the test teams;

– allow updates to requirements as needed. These updates will impact the tests (design of test cases and scenarios, test data, test levels, etc.) in addition to the components to be modified.

Depending on the business domains and the criticality of the test levels, the documents must demonstrate a level of independence between the authors and the verifiers as required in the standards applicable to this domain (e.g. DO178B/C for the aeronautical).

7.4. Test policy

As mentioned in Chapter 3, the test policy succinctly defines the objectives and principles of testing within an organization. It defines what should be achieved by the tests, but not how the organization should perform the tests. The test policy provides a framework for establishing, reviewing and continually improving the organization's test policy. It applies to internal fulfillment activities of the organization and can be used for activities that the organization performs as subcontractors for customers. It may not apply to all of the organization's projects, components, products or systems. For example, Research and Development projects or projects with a budget below a certain value may be outside the scope of this testing policy.

A typical test policy shall include:

– Document specific information:

  - unique document identifier;
  - issuing organization;
  - approving authority;
  - history of changes.

– Introduction:

  - perimeter and scope;

- internal and external references;
- glossaries.

– Statements:

- test objectives;
- testing process;
- organizational structure of the tests;
- education, training and ethics of testers;
- standards and other applicable policies;
- measurement of the value of the tests;
- archiving and reuse of test products;
- continuous improvement of testing processes.

It is necessary to ensure that the elements defined in the organization's test policy are correctly implemented by co-contractors and sub-contractors. This may involve making available to them elements of the organization's repository, periodic metrics and measurements, etc.

7.5. Test strategy

We detailed the testing strategy in a previous chapter. Here are the important things to remember in this document:

– test objectives definition;

– impact of development cycles and in the event of use of the agility of the postponement of stories to a next iteration (and their impact on end-to-end operation);

– approval by stakeholders;

– management of developments, versions, components (including hardware), regression tests on each level, test data, etc.

– definition and interaction of test levels with each other.

It is important to highlight that the "Test Strategy" document should highlight the logistical aspects associated with achieving the test objectives.

7.6. Master test plan (MTP)

The master test plan (MTP) defines the elements common to all test levels and the elements common to all systems composing the system-of-systems. It can also include:

– all the elements common to the project;

– origin of the specifications and requirements, and how these will be verified and validated;

– test techniques to implement according to the level of criticality;

– level of detail of the test deliverables;

– the test objectives by level (i.e. the success criteria for project and product);

– how to break down the tests to consider:

- verification of the absence of regressions in the applications interacting with the new components,

- verification of the life cycle of data exchanged between systems, regardless of the original system (legacy or new),

- verification of the operation of new systems,

- end-to-end system-of-systems verification,

- archiving, monitoring and traceability of changes, etc.;

– evidence to be provided, both as entry criterion and as exit criterion for each process;

– how the processes will be verified, and their deliverables validated;

– etc.

As the master test plan develops the objectives defined in the strategy, it may be necessary to recall the applicable objectives on the project.

If the project is mainly sequential, each test level should be properly named and its content – its objectives – should be clear to each participant. The scope must be understood and consistent with other levels.

If the project is mainly Agile, the objectives of each group of sprints must be defined in order to clearly identify the milestones and meeting points between the various sub-projects/teams/deliverables, as well as the level of quality and completeness necessary.

It is not necessary to modify the "customer" terminologies of the test levels or phases as this frequently leads to rejections. However, if the terminologies are confusing, ambiguous or contradictory, the presence of a glossary makes it easier to understand.

Test phases for systems-of-systems frequently include a "master test plan". This defines the test levels and the test organization, with test objectives for each test level.

Having one or more test environments per level avoids side effects during the execution of test campaigns:

– a pre-test or smoke-test environment makes it possible to verify that the software provided meets the entry criteria for the test level;

– a development or integration environment allows developers to reproduce anomalies if necessary;

– one – or several – level test environment(s) will make it possible to verify and validate the requirements by manually or automatically executing the planned test cases.

The management (definition, implementation and validation) of the level test environment must be carried out taking into account the moment when this environment will be used. For example, if we want to automate the tests, a finalized environment must be available to allow the recording of the automated tests, or at least the linking of the real objects with the objects defined during the tests. The absence of such an environment early enough will lead to delays in the execution of the tests (inability to verify the proper functioning of the test cases on the test execution environment).

Activities described in the Master Test Plan should consider that the test execution phases identify defects and will need to be re-executed again, after corrections have been delivered.

Depending on the test levels, the level of formality of the test execution will vary from very informal to very formal when approaching the delivery of the final software. Similarly, reporting frequency will often increase as the delivery date approaches.

Since systems-of-systems are generally composed of multiple components, and these components must be validated, it is necessary to ensure that the validation or verification criteria (DOD – Definition of Done) have been correctly applied. This requires a formal test phase, on a controlled environment, with formalized tests and identified test data, as well as an element correctly identified in configuration. This requires a formal phase that ensures that all the prerequisites for the execution of these tests – generally acceptance tests – are present. Such a phase is materialized by a meeting (TRR – Test Readiness Review) which will formally ensure that the prerequisites are present and that the tests can begin. A TRR must identify:

– components delivered and to be tested;

– test cases to be executed, manual or automated;

– requirements to be validated by these tests;

– test data to be used;

– expected results that will serve as a basis for comparison of results obtained.

At the end of this period of formal tests, the results will be evaluated and another formal meeting, the Test Review Board (TRB), will decide on the successful passing of the milestone. In the event that anomalies have been identified, the TRB will have to decide on how to deal with these anomalies: either they will be accepted as limitations, or they will have to be corrected or the delivery will have to be postponed to include the correction of these anomalies.

TRB makes it possible – formally – to accept the end of a test level, often an acceptance test level, within the framework of a system-of-systems. By signing the meeting minutes, all members of the TRB commit to the quality of the tests carried out and of the component.

NOTE.– The term "test plan" is often used by test tool vendors to reflect a list of tests.

7.7. Level test plan

Level test plans define the specific testing activities for a given level. The master test plan template can be used, adapting to the specifics of the level in question, or you can use a specific template.

If we refer to the ISO/IEEE29119-3 (2010) standard, the test plan can contain the following modules:

– document-specific information, such as unique identifier, issuing organization, approval authority, change history;

– an introductory module, including the scope, references and glossary of terms used;

– an introductory context module with information on the project, the components to be tested, the planned scope of the level tests, the assumptions and constraints as well as the stakeholders;

– a module on exchanges and communications between teams;

– a module on risks, with product risks and project risks;

– a section on test strategy implementation with test phases and test types, test techniques to be used, end criteria, test environment and data requirements, tools and deliverables;

– a module on the estimation of test loads and tasks;

– a module on staffing (responsibilities, hiring and training), responsibilities and tasks;

– a module on schedule, with the criteria for suspension and resumption;

– a reference to the higher level risks and how they are addressed, and any information justifying deviations from the test strategy and/or master test plan.

7.8. Test design documents

The term "Test Design Document" is used here generically to reflect all the documents that will be necessary to cover a quality characteristic, a risk or a series of associated risks, or even a set of functionalities that can be grouped together, within one test level or transversally over several test levels.

Therefore, we could have a test folder to cover performance testing and another test folder for security aspects. Similarly, we could have multiple test folders to cover multiple functional areas.

As a result, in the test files we could have a description of the static tests that will be considered and the dynamic tests that will have to be executed.

For static testing, we will have review checklists as well as static analysis rules.

With respect to dynamic testing, we will include what refers to specification of test cases and test scenarios or procedures, aspects of test data and test environments (initialization, saving, use and cleanup or restoration data and test environments).

7.9. Test case specification

The test case specification document describes the details of each test case covering the component, product or system. The level of detail of this description will vary according to the level of knowledge of the people in charge of the execution of the tests and the need or not to automate the test cases: if the people in charge of the execution of the tests are specialized in the business domain, it may be useless to describe the test case too finely; on the contrary, if the people are not experienced in the field or if it is necessary to automate these tests, it may be necessary to detail these test case descriptions. This will reduce uncertainties during test execution.

More and more tools make it possible to describe test cases and automatically link them to risks or specifications. It is therefore important to ensure – if the risks or specifications change – that the version information is associated with the test descriptions.

7.10. Test procedure specification

Test cases can be grouped into test scenarios or test procedures. The proposed content is as follows:

Test sets:

i) unique identifier;

ii) objectives;

iii) priority;

iv) contents.

Test procedures:

i) unique identifier;

ii) objectives;

iii) priority;

iv) start up activities;

v) test case to execute;

vi) link to other procedures;

vii) stop and finalize.

7.11. Test data specifications

The test data, both the input data and the data to which the obtained data will be compared, must be properly identified and referenced:

i) unique identifier;

ii) description;

iii) responsibility;

iv) period needed;

v) reset (reinitialization) needs;

vi) archiving or removal.

7.12. Test environment specification

The test environments, whatever the level of test where they are used, must also be correctly identified and referenced:

i) unique identifier;

ii) description;

iii) responsibility;

iv) period needed.

7.13. Reporting and progress reports

The purpose of reporting is to trace data in order to estimate the progress of the implementation of activities in relation to an objective. The data reported must allow informed decisions to be made on the project. This information covers the following:

– progress with regard to achieving target objectives;

– use of resources to date and the rest to be done until the objectives are reached;

– compliance to deadlines, delays and their causes, remaining durations of activities;

– level of quality of the objects (components, software, systems, functionalities, etc.);

– level of quality of the processes implemented.

Key Progress Indicators (KPIs) will therefore have to be defined to allow the monitoring of this information.

Remember that reporting provides value only if it is correctly understood by the parties to whom it is addressed. This implies that it must be easily understandable and unambiguous to allow decision-making on the project, comparison of the project with other similar projects and evaluation of the achievement of project objectives.

7.14. Project documentation

Test project documentation is divided into two groups: test input documentation and test output documentation. This applies also to systems-of-systems projects.

Input documentation for test activities includes among others:

– contract;

– specifications or list of requirements (product backlog in Agile);

– hardware and/or software architecture;

– definition of user interfaces (UI or GUI);

– specifications of interfaces with other applications;

– architecture specifications;

– general, functional and technical specifications;

– detailed specifications (for each component or system);

– user stories (for Agile developments);

– standards, norms and applicable "best practices";

Output test documentation includes among others:

– test policy;

– test strategy;

– master test plan (MTP) and level test plan(s) (LTP);

– product risk analysis (ARP);

– test design documentation;

– test conditions;

– test data;

– test results;

– defect reports;

– interim test reports;

– final test report;

Those documents may be present during a system-of-systems project and used by test teams.

7.15. Other deliverables

Depending on the projects and organizations, many other deliverables may be required. Among the most important, we have the following:

– The Compliance Creation Plan (CCP) which determines, for each of the requirements – functional or not – present in the contractual and applicable documents, how compliance will be determined. In general, this compliance is determined by Inspections, Analysis, Demonstrations (i.e. tests in dedicated environments) or Acceptance Tests (in a real production environment). This CCP will make it possible to determine the review and analysis actions that will have to be carried out in the design phase, and therefore to predict the content of the deliverables of the design milestones (PDR and CDR).

– The Requirements Traceability Matrix (RTM) which makes it possible to make the link between (a) each of the requirements – functional or not – present in the contractual and applicable documents (e.g. standards, and also the CCP for actions of the type "D") and (b) the test evidence provided during the execution of the tests on the delivered software, equipment, subsystems and systems.

8

Reporting

Reporting provides information on the current state of the project against a desired state defined during test planning, to make informed decisions about the project. The data reported must be adapted to the stakeholders, both in terms of level of detail and in terms of frequency, in order to allow them a quick and clear understanding, and sound decision-making.

8.1. Introduction

Throughout the test project, or even its various iterations (sprints, release trains, etc.), test campaigns and versions of the components, products or systems to be tested, the test manager must provide information on the progress of the test activities, the quality level of the product, and the optimal use of the resources allocated to it to carry out the tests.

Many questions will arise, such as (non-exhaustive list):

– What is the current quality of the product?

– Which components, products or systems will need more testing, where are there more defects and therefore where should we focus our efforts?

– What is the coverage achieved (components covered, functionalities available)?

– How efficient are the testing processes?

- What is the current progress, and is it in line with initial forecasts?

- How do you measure the effectiveness of your testers, components, products or systems under test?

- How do you evaluate that the level of quality achieved for the components, products or systems under test corresponds to what is expected by customers?

- How do you ascertain that the planned activities are carried out correctly and that your teams find all the defects?

- How much will these testing activities cost? How can these costs be reduced (without reducing quality)?

– When will the test campaigns be completed? How to reduce the duration?

– What are the remaining risks?

All these questions will be asked by your customers and the business managers you meet. In general, the areas will be the following:

– Level of quality of the components, products or systems subjected to the tests, and how to improve it?

– Resources spent (cost to date) and remaining to be spent (cost to completion) to carry out the tests, and how to reduce them while maintaining the same level of quality?

– Duration of test activities (planned, carried out and remaining to be done) to ensure the required level of quality, and how to reduce them while maintaining the same level of quality?

– Risks – financial, technical or other – to which the company is exposed if the components, products or systems go to the next stage (e.g. production)?

The people who will ask you these questions can be grouped into various groups: hierarchy, customers and users, developers and design teams, test teams, support and operations teams.

It is obvious that the periodicity of the reporting that you will carry out will evolve according to the phases of the project. At the start you will need to define the strategy, resource and time requirements, dependencies on other elements, etc. On the contrary, when running the tests for each iteration, sprint or level, the frequency will increase, until it is weekly or daily.

The reporting is therefore based on measurements and compares the data planned before the execution of the activities, with the data reported following the execution of the activities. This comparison can be made on the basis of tables, but it is often more understandable to visualize it in the form of graphs in order to be able to visualize trends.

Warning: data – measurements – reported represent a level of information at a moment in time. Test projects, especially those related to systems-of-systems, are fluid and continuously evolving. As a result, the metrics are not always fully representative.

8.2. Stakeholders

As mentioned earlier, stakeholders, the groups of people who will hold you to account, can be grouped into four categories. Each of these categories has stakes or interests that may diverge from the interests of the other parties:

– hierarchy, program or project management team, financial and general management;

– customers, business representatives and users who will use the product or system;

– developers, including co-contractors and sub-contractors, managers of related projects etc.;

– test teams (analysts, automation engineers, test designers and executors).

In addition to those stakeholders directly involved in testing, other stakeholders may be interested in the test results, such as the support and operations teams who provide the testing support and the logistics necessary for the operation of the tests. Components, products or systems that make up the system-of-systems. We will have, for example:

– the people in charge of deploying the components and products developed on the various environments;

– the teams responsible for the administration and day-to-day management of components, products and systems;

– the people who will carry out the transfer and deployment of the developed products to the production environments and systems.

These stakeholders will want to know:

– What components should be transferred, where are they, what is the correct version, where are the proofs that the component is of good quality, what are the administrative activities to be carried out (creation of user profiles, configuration of accounts and data, etc.), which flows should be opened and monitored, how can we ensure that the components are working correctly, where is the administration and maintenance documentation, etc.?

– The needs in terms of hardware and operating systems, deployment framework (e.g. AppStore or Android store if certain components are available on mobile devices), versions of languages or databases, systems ERP, CRM or others to configure and connect to the system-of-systems, etc.

– The activities to be implemented to deploy the version of the system-of-system, to ensure that each of the components connects correctly, that the data is correctly exchanged, and that monitoring information makes it possible to ensure the good operation of every system of the systems-of-systems. Of course, you will also have to make sure that you can roll back without losing data if unfortunately a deployment does not go well.

In the support category, we will also include the teams in charge of user and customer training, the creation of online courses, installation and user manuals, etc.

8.3. Product quality

The quality of a component, product or system can be measured by comparing the requirements that are covered against the total requirements defined. This can be done using traceability matrices from the requirements – or use cases or features – to the test cases that verify them. However, many requirements are not clearly identified or listed, so this level of quality may be incomplete. The percentage coverage of product requirements and/or quality characteristics may need to be weighted according to the priority of these requirements and/or quality characteristics.

On the contrary, the percentage of requirements coverage should take into account the impact on the end-to-end functioning of the system-of-system: if a single component in a chain does not work, the entire chain does not work. This aspect of end-to-end operation is difficult to take into account in the absence of an exhaustive mapping of the components, their versions and their interactions with each other. The full combinatorics of system-of-systems functionality will probably be too large for all combinations to be tested, but a subset – the size of which depends on the criticality of the system-of-systems – should be able to represent a sample of sufficient test cases (smoke test type).

8.4. Cost of defects

One way to estimate the costs of anomalies is to determine the cost of detecting and correcting them based on when they are discovered.

NOTE.– Like any statistic, this measurement depends on the number of measurement points. When the number of defects is small, the sum of the costs of the quality improvement measures is to be divided by a small number of defects, leading to a large number, while the quality is in fact better.

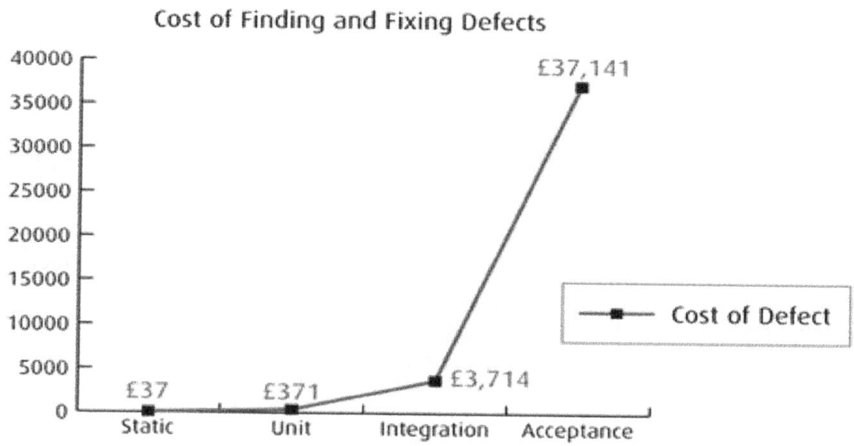

Figure 8.1. *Cost of fixing defects*

Another way to measure the cost of defects is to estimate the impacts of anomalies on customers and users. In a system-of-systems, the impact of an anomaly can be disproportionate to the cost of correcting this anomaly. It suffices to take as an example defects such as those which led to the loss of the first Ariane 5 flight, or the immobilization of fleets of Boeing 737-800 Max aircraft. The amounts defined in Figure 8.1 come from information published some years ago, and must certainly be updated to fit your own organization.

You could also measure the costs and added value of testing in terms of:

– return on investment;

– detection costs;

– internal and external correction costs (these external costs may be huge with embedded software);

– added value of the teams in terms of defect prevention;

– improved effectiveness and efficiency of test teams;

– increase in the quality of decisions by providing accurate information.

> *Attention*: using the cost of defects will give incorrect results at the limits. For example, if there is only one defect left, then all the quality expenses add up to make a prohibitive cost when there is, in fact, only a very small number (one) of defects to correct and the system-of-systems is of good quality.

8.5. Frequency of reporting

The effort to generate reliable reporting is significant. This may involve performing exports from data (e.g. exporting information on defects, on the execution of tests, etc.), processing the exported data (e.g. grouping defects according to their criticality or the components to which they apply, associate test executions with component versions, etc.) or generate reporting supports (e.g. progress charts, trend analysis, consolidation of data tables, etc.). It is sometimes necessary to do similar work. Reporting of tests and test results will vary in frequency depending on progress. The closer the project is to important milestones, the more the project management will want to have up-to-date data.

8.6. Test progress and interpretation

The progress of testing can be measured in various ways, according to different views:

– percentage of requirements covered (planned vs. achieved);

– percentage of risk coverage (planned vs. current level of risk);

– percentage of coverage of components, products or systems;

– percentage of test case coverage (planned vs. actual);

– percentage of resource utilization (planned vs. actual).

Regardless of the mode of representation of advancement chosen, it is important to note that everyone will seek to present a level of advancement that highlights themselves even though the usefulness is not demonstrated.

For example, measuring progress that compares the actual to the planned is useful "all other things being equal". It is indeed important that progress be measured on elements that are of real use. Some requirements are critical while others – in greater numbers – are not, so a measurement of the coverage of requirements by designed tests must also focus on the criticality of these requirements. If all the test cases focus on the non-critical requirements, we will get a sense of satisfaction that is not needed because the critical requirements are not properly covered. Similarly, if the design phase of the test cases and scenarios did not create many tests, boasting of 100% execution of the designed tests is of no use.

The test manager must therefore simultaneously monitor several indicators covering at the same time:

– the requirements and their priorities or criticality. It is necessary to process – design and execute – tests that cover the highest priority and criticality requirements, as long as the components that implement them are available;

– the components (+ products and systems) and the requirements they meet, as well as the risks they minimize, the size and complexity of these components as well as their criticality; this implies that the Test Manager must ensure that the elements (e.g. requirements or software components) are delivered to the tests on time and in such a way as to allow the good progress of the test activities;

– the risks and how they are minimized in terms of percentage of occurrence or impact;

– the test cases planned (or to be planned) as well as their number, according to the criticalities to be covered (the higher the criticality, the more test cases must be planned, with different techniques to find as many defects as possible);

– the actual use of resources (personnel, time, etc.) compared to the planned optimal use. This makes it possible to anticipate load changes.

8.6.1. *Requirements coverage*

The objective here is to demonstrate the coverage of the requirements, according to their priority, by tests and then by successful executions of these tests. According to the TBX methodology, the requirements will be covered according to a decreasing level of priority. The higher the requirement, the sooner it should be covered and the more testing techniques should be used.

Coverage of requirements must be done in parallel with coverage of risks and priorities, each requirement being associated with a level of risk and a level of priority.

Coverage of requirements, as well as the follow-up of their evolutions, is an activity that continues throughout the project because the requirements are likely to change throughout the project. Any change – addition, modification or deletion – of requirements may impact the tests to be performed to verify their proper functioning. With each requirement modification, it will be necessary to propagate the information to the development teams and to the test teams. The test teams will modify the test cases and scenarios according to the modified requirements (adding, modifying, deleting test cases and test scenarios). It may be necessary to see the impacts on other components, products or systems as well as on the interfaces and ensure the availability of the necessary test data.

Coverage of requirements is aided by the use of requirements management tools that include configuration (or version) management making it possible to identify changes in the versions of components, products and systems, and also to link these versions to test run results for each of the specified versions.

A quick measure of the rate of coverage of requirements by tests is to ensure that each requirement is covered by – at least – one test case. Depending on the level of criticality, priority or risk of each requirement, more testing will be required. It is possible to cover several requirements with a single test scenario (e.g. a use case covering several test cases), so it is the test cases that must be measured and not the scenarios or procedures of test.

8.6.2. *Risk coverage*

If we have decided to group the risks by risk categories (see risk calculation), we can report the risks according to these categories. We will then focus on the evolution over time of each of these categories.

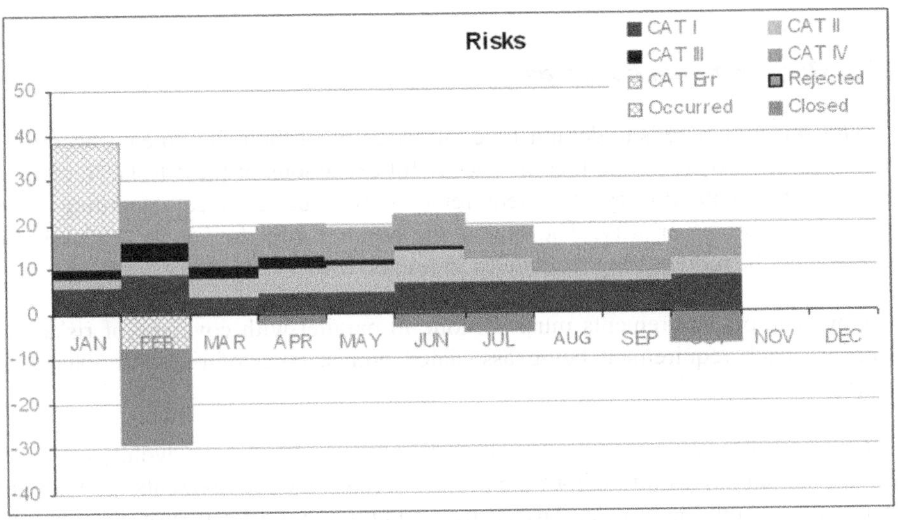

Figure 8.2. *Evolution of risks by risk category. For a color version of this figure, see www.iste.co.uk/homès/systems1.zip*

In Figure 8.2, we can see that the CAT I risk category is increasing during the months of March to October, without showing any decrease. On the contrary, the CAT III category decreases between February and June to disappear in July. It can be concluded that the risks of the most important categories are not treated while those of the less important categories are correctly mitigated.

8.6.2.1. *Top 10 risks*

Given the often large number of risks associated with a system-of-systems and the impossibility of periodically analyzing each of the risks, detailed reporting on the five or ten most important risks depending on the area should be addressed. This does not release the test manager from analyzing all the risks, to prevent less important risks from becoming more important.

8.6.2.2. *Project comparison*

Project managers have different sensitivities regarding risk. Some individuals will accept risks more readily than other individuals. It is therefore possible to compare similar projects according to the state of risk.

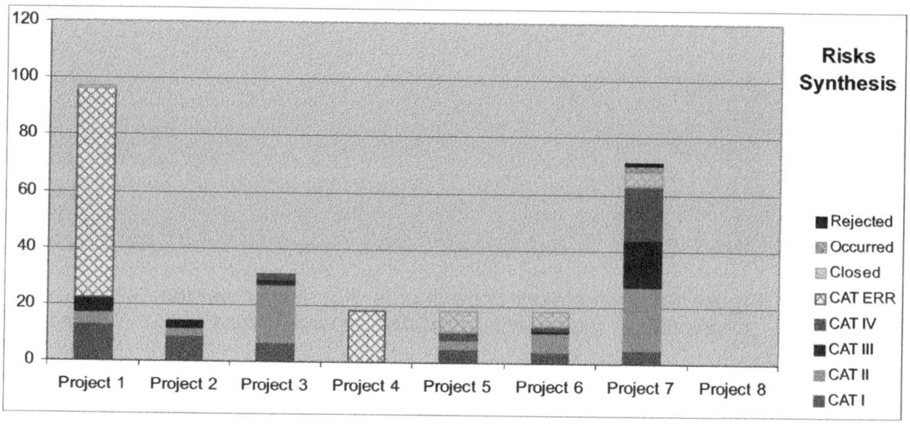

Figure 8.3. *Comparison of risks by projects. For a color version of this figure, see www.iste.co.uk/homès/systems1.zip*

Figure 8.3 allows us to compare projects of similar size and purpose and to see the difference in risk sensitivity between projects 5 and 6, on the one hand, and project 7, on the other hand. Even though the risks of category 1 are of similar volume, the risks of the three other categories are clearly more numerous for this project 7.

8.6.2.3. *Comparison of components*

Another way of estimating risks is related to the number of defects identified in each component (software or hardware). We can, from the analysis of anomalies identified on the project, identify the component or components with the most anomalies. According to the main principles of the test, these components will statistically have more defects than others. Of course, the size of the components (in lines of code, function points or complexity) will come into play.

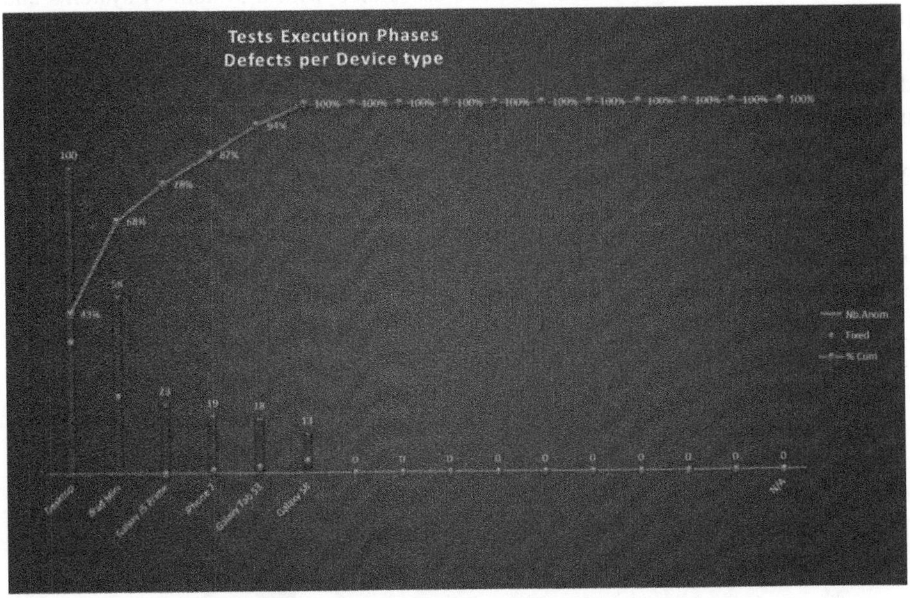

Figure 8.4. *Defects per components. For a color version of this figure, see www.iste.co.uk/homès/systems1.zip*

Figure 8.4 shows that the majority (68%) of faults are identified on two components. Another important piece of information is the percentage of bugs corrected: the bugs for the two components on the left are being processed (about 40% of the bugs for each of these components) while the number of bug fixes is significantly lower for each of the four other components.

8.6.3. *Component or functional coverage*

We can report components or functionalities based on a grouping of anomalies according to these criteria. We sometimes find an application of Paretto's law (20% of the components contain 80% of the anomalies).

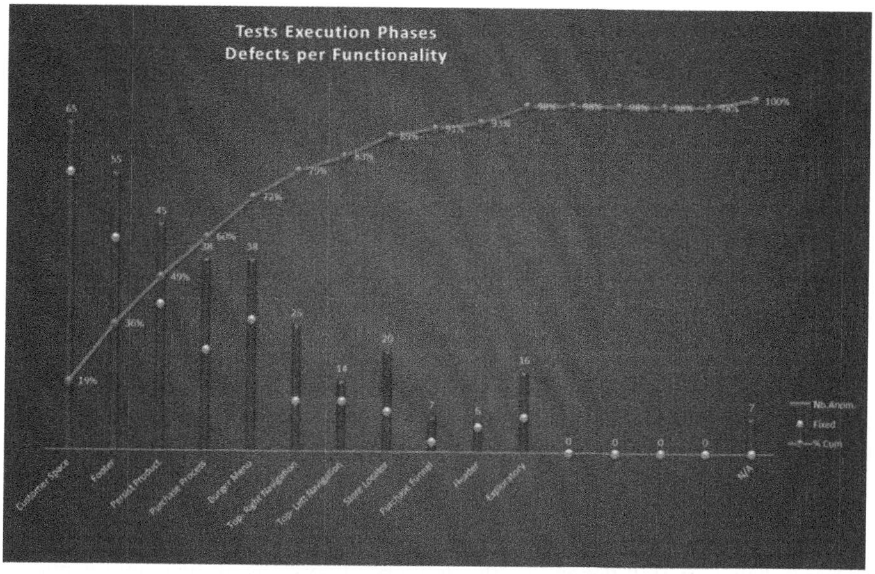

Figure 8.5. *Coverage of component or functionality. For a color version of this figure, see www.iste.co.uk/homès/systems1.zip*

If each component or feature is managed by a separate team, and provided that the components or features are of similar volume, it is possible to identify the components or features most prone to defects.

If we look at the number of defects corrected (or the percentage of defects corrected), it is also possible to identify the most reactive teams in the correction of these anomalies.

According to Jones (2017), some components – often less than 5% of the components within large systems – have a large number of anomalies – more than 50% of the total anomalies. These modules (called EPM for Error Prone Modules) should be removed and rewritten. According to IBM, on the IMS database application comprising about 425 modules, 57% of the defects were concentrated on a subset of 31 modules (7.4%). These modules had a high cyclomatic count and did not use inspections.

8.7. Progress and defects

During the execution of test campaigns, it is important to measure the evolution of anomalies, their detection and their correction. This will identify:

– the ability of test teams to discover new defects;

– the capacity of the development teams to correct the defects discovered and the average time for fixing;

– the number of anomalies remaining to be corrected (defect backlog);

– the number of times a defect has been reopened. This indicates a concern in the ability to properly fix a defect.

8.7.1. *Defect identification*

The purpose of testing is to discover defects so that they can be corrected. When the component is provided in a version that would correct this defect, it must be retested in order to confirm the correction of the defect. It is therefore a process that takes place over time, and whose objective is to end up – in the long term – with a component free of defects.

Defect discovery will depend on the delivery of versions of the components to be tested (some may include more defects), and the workload assigned to the execution of the tests (number of people performing tests and hours of testing per nobody).

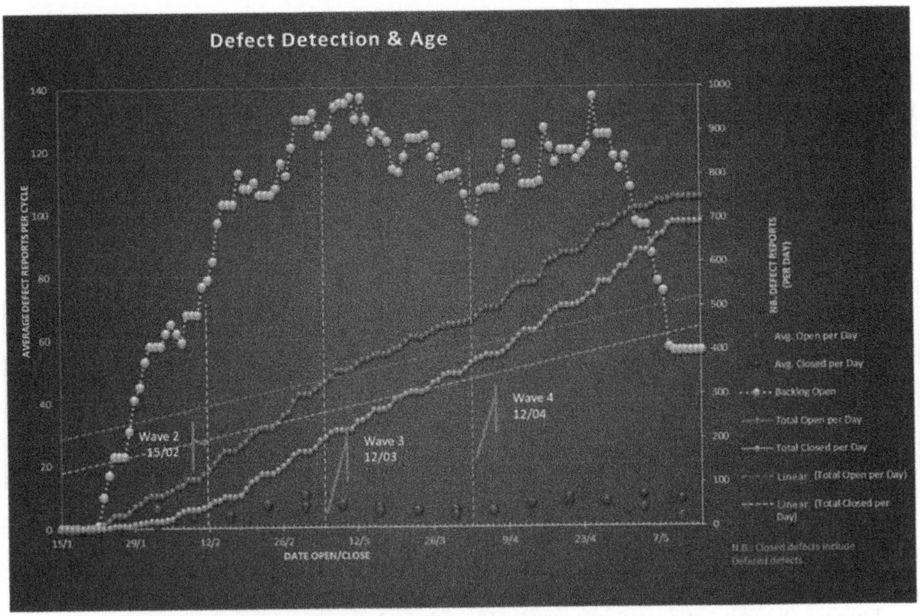

Figure 8.6. *Defect discovery and correction. For a color version of this figure, see www.iste.co.uk/homès/systems1.zip*

In Figure 8.6, we can visualize in the lower part the number of defects identified and corrected per week. The assumptions are (1) stability of the test execution team's workload and (2) stability of the application and the execution environment.

The number of defects identified (red dots at the bottom of the graph) is generally greater than the number of defects fixed (green dots). This is reflected in the evolution of the defect backlog curve (orange curve).

If we look at the cumulative evolution of the defect detected (red "Total Open per Day" curve), we see stability until the end of April, then a trend towards stability. We can conclude that the test team finds less and less defects in the component. The main assumption is that the testing team continues to test the application with the same workload.

8.7.2. *Defects fixing*

In Figure 8.6, if we look at the cumulative evolution of the number of anomalies corrected (green "Total Closed per Day" curve), we note that this curve, after a relatively long period where it increases in the same way as the anomaly detection curve, tends to join it in the last month of the chart. The conclusion we can draw here is an improvement in the quality of the component.

Two additional pieces of information we can glean from comparing the cumulative anomaly identification and remediation curves are as follows:

– At a given instant, the vertical difference between the two curves gives the number of uncorrected open defects.

– For a given number of defects, the horizontal distance between the two curves gives the average duration of fixing these defects.

8.7.3. *Defect backlog*

In Figure 8.6, the "Backlog Open" curve identifies the number of anomalies remaining to be processed. Putting this number in relation to the average number of bugs fixed per day allows us to determine whether – all other things being equal – the teams in charge of bug fixing will be able to deliver the fixes on time.

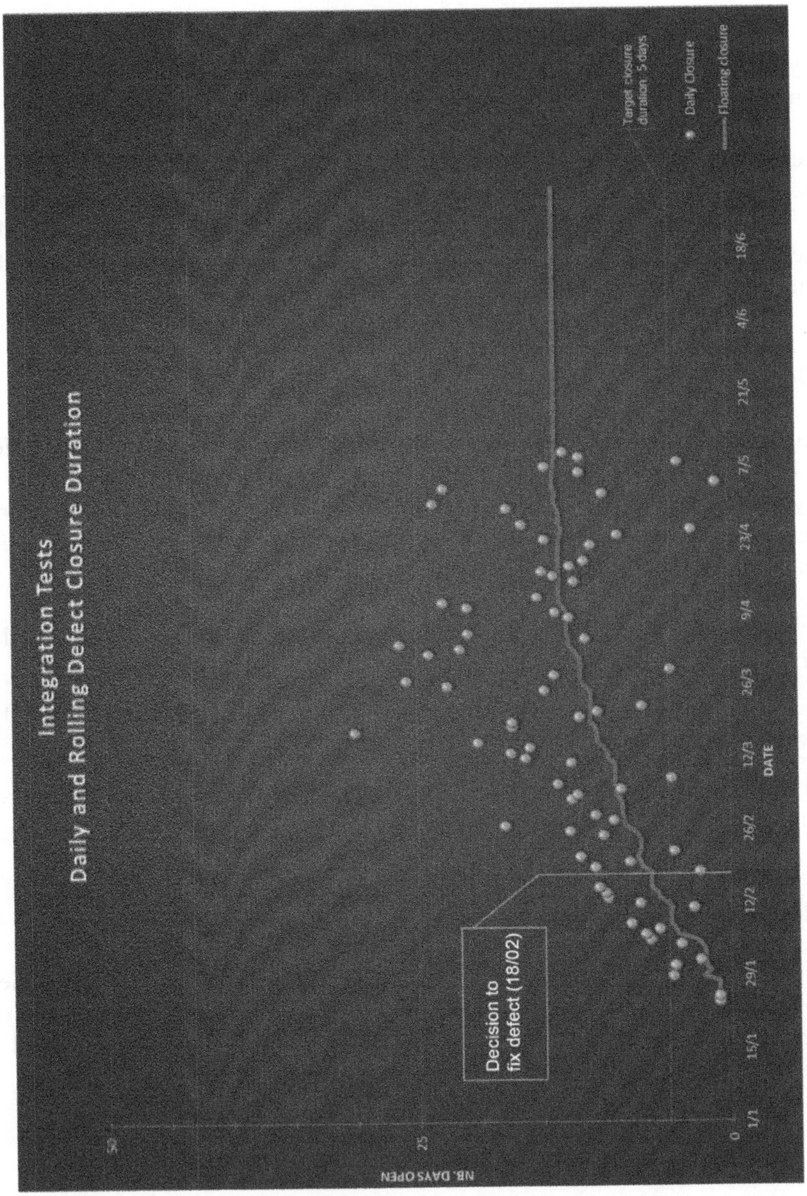

Figure 8.7. *Average and cumulative defect fix duration. For a color version of this figure, see www.iste.co.uk/homès/systems1.zip*

We can visualize the average defect remediation time with Figure 8.7. If the average target duration of correction is five days (green line), and the average daily duration is represented by the yellow dots, the cumulative curve representing the average cumulative duration of correction clearly indicates that we rather have an average duration of 1–15 days between the identification of an anomaly and its correction.

8.7.4. *Number of reopened defects*

The normal process for a defect is that it is discovered, sent for correction, corrected and re-delivered to testing for validation and then closed. Any referral from the development team to the test team (number of re-openings of an anomaly) therefore indicates a problem. The problem may be, for example, (1) an inability of the test team to clearly explain the anomaly or (2) an inability of the development team to correctly fix these anomalies without introducing new anomalies.

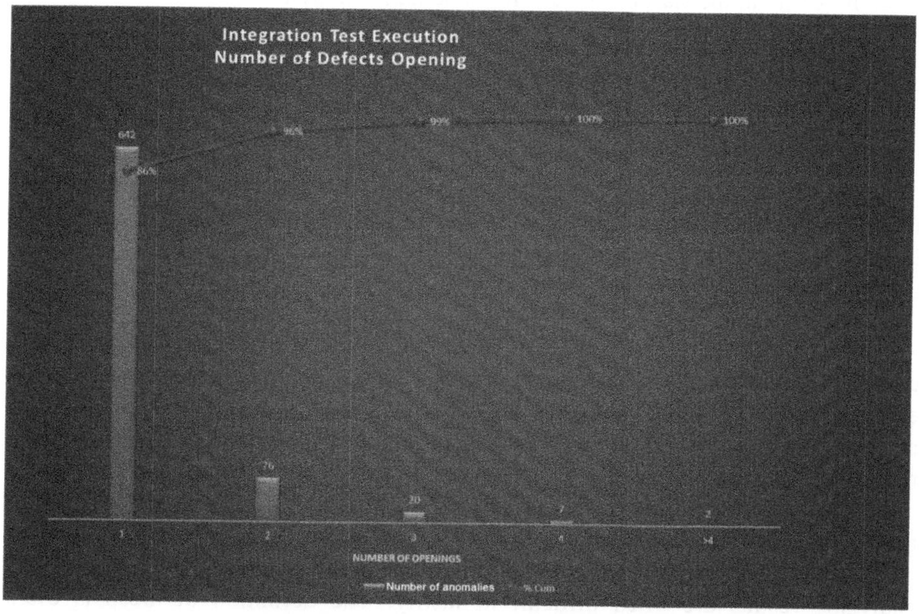

Figure 8.8. *Number of defect openings*

Consider that in general any opening (and reopening) of a defect costs between 2 and 4 hours of work. In Figure 8.8, we see that 76 defects were opened twice, 20 defects were opened three times, seven defects were opened four times, and two defects were opened more than four times. Counting 4 hours per opening, we therefore have 580 hours of work lost as a result (i.e. more than 72 days of work).

8.8. Efficiency and effectiveness of test activities

There are numerous ways to answer these questions. You could, for example:

– Measure the coverage rate achieved, whether for the coverage of User Stories by test conditions or for the coverage of test cases by automated scripts, or for the execution of tests on the current version. This information is useful for checking progress but will not provide you with information on defects.

– Check the number of test cases planned per requirement or per User Story to ensure that they are correctly covered. This will allow you to ensure minimum coverage (usually nominal operation) but will not allow you to identify exception cases or borderline cases. Such a measure is useful to ensure that there is a sufficient level of quality at the start of the test implementation activities.

– Compare defect detection and correction curves, to ensure that the backlog of uncorrected defects remains manageable and does not grow too large. An increase in the default backlog will tend to increase technical debt as more defaults will be pushed to a subsequent delivery.

– Identify the most effective testing techniques (finding the most anomalies) or the most efficient (finding defects with the least effort). This makes it possible to identify the most used and/or effective test techniques – dynamic and static – with the aim of generalizing them.

Effectiveness and efficiency are difficult to measure and report. Some propose to compare – via burn-up or burn-down graphs – the results obtained compared to the expected results, whether we are talking about workloads ("man days"), deadlines or number of actions carried out (test cases executed, anomalies detected, corrected and retested, etc.). Each of these measures is likely to be impacted by bad initial objectives (too or not ambitious enough) and generally only reflects a small part of the picture. One solution would be to have a comprehensive measurement policy in place for all aspects of testing, but the workloads associated with measuring and analyzing results are often difficult to justify. The definition of areas is often implemented for continuous improvement.

We see that it is important to have several lines of vision and reflection to properly cover the efficiency (profitability) and effectiveness of testing activities.

8.9. Continuous improvement

Implementing a continuous improvement process allows you to continue to improve over time. This involves estimating the current level of performance (with KPIs), the level to be achieved (with the same KPIs), the improvement actions to be put in place and the deadlines for obtaining the results (therefore measuring again based on the KPIs defined before). In the context of a system-of-systems, given the duration and complexity, it is essential to ensure that the quality of the processes is constantly improving.

Continuous improvements must be measurable – that is, they must be associated with metrics – and be measured over time to identify whether improvements are – or are not – being achieved. In addition, the choice of areas for continuous improvement should follow an identification of the improvement points identified during CTP (Critical Test Processes) and TMMi (Test Maturity Model) type audits and be prioritized according to the expected benefit of each proposed improvements.

In the context of systems-of-systems, the parts of the organizations not directly impacted by the project may also wish to benefit from feedback and continuous improvements.

8.9.1. *Implementing continuous improvements*

It is important to understand that it is illusory to impose improvements without the support of those who will implement them. Indeed, the usual reaction when something is imposed on us is to refuse to do it (remember your reaction when your parent asked you to clean your room). On the contrary, if the improvements are proposed by those who will implement them, the implementation effort will be carried out more quickly and with more enthusiasm.

Awareness by all actors is therefore necessary and the elements for improvement must be identified. They can relate to improving the relevance of reporting, and also to each of the aspects of design, production, verification and validation of achievements (design, development, integration, etc.).

We position in this chapter the treatment of continuous improvements, and you understand that these methods of continuous improvement will apply to other aspects than simple reporting.

8.9.1.1. *Identification of improvement points*

Identifying areas for improvement comes easily by interviewing people who receive deliverables and are dissatisfied with their level of quality, delays in delivery dates or other similar aspects. Identification can also come from recriminations or complaints from co-contractors, sub-contractors or customers, just as it can come from feedback from test or quality assurance teams at various levels of test according to your system-of-systems organization.

When you have identified points for improvement – in fact you first identify "pain points" or points where you are in pain – you must identify the probable root causes. One way is to use the so-called "five whys" method. The principle is to ask why the quality is not suitable, to find several causes, and then to dig – five times – with a series of "Why?" questions. You should thus identify the real causes at the origin of the problem.

Attention:

– there can be several causes that come into play;

– keep a focus on what is statistically representative, so causes identified will have a greater impact;

– consider organizational, political and relational causes, including personal conflicts (this list is not exhaustive) because these elements are often forgotten but have an important effect.

As soon as the root causes have been identified, you will have to implement improvements according to the necessary investment, the envisaged gain, the timeframes for implementation and obtaining the gains. You will be able to define objectives that are "quick wins" and others that will require more substantial investments.

8.9.1.2. *Measurement of improvement objectives, KPIs*

As with anything, continuous improvement is a process: you need to know where you are starting from and where you want to go. This involves measuring the current state, with metrics, and checking – throughout the process – if the planned objectives are achieved (with the same metrics).

We could improve the quality of new code or code already produced (Legacy code). This involves measuring the level of quality of these two types of code, which can be done via metrics based on the defects identified:

– for the quality of Legacy code, we could measure the percentage of "regression" type defects on parts of code that worked correctly in the past;

– for the quality of new code, we could focus on defects related to new User Stories or recently created code.

These two types of code will allow us to decide on which objectives to focus improvement efforts.

One solution would be to use the GQM method, detailed in section 16.1 of volume 2.

8.9.1.3. *Definition of improvement activities*

Defining improvement actions can benefit from suggestions from other stakeholders (e.g. Development, Business, but also test team, etc.) in order to establish a feeling of belonging and collaboration: everyone will feel involved and enhanced by the use of solutions offered. In the definition of improvement actions, you must consider the return on investment (how much the implementation of improvement actions will cost you and how much they should bring in) in order to limit as much as possible, the impact, slowdown or increase in workload, team members. Other elements to consider: the initial measurement – and the following ones – and the conclusions that one plans to take based on the measurements. All these metrics will make it possible to define KPIs (Key Process Indicators, Key Indicators).

It is indeed important to know what is the current state of effectiveness of the testing or quality assurance actions that are carried out, to have a starting point to measure the improvements. Another important element is to define – from the start – how the measured values will be interpreted, to coordinate them with the axes of improvement defined previously.

8.9.1.4. *Measurement of improvements*

Once the KPIs and improvement activities have been defined, it is necessary to ensure that the latter are effective. This means ensuring that they are formalized, understood, implemented and bring the expected benefits. It is not clear that the expected benefits are directly identifiable, so intermediate KPIs may need to be considered and measured.

Other points to consider to obtain correct measurements are:

– To ensure that the objectives of the KPIs and measures are correctly understood by those responsible for carrying out the measures. This ensures that the purpose of the KPIs and the importance of the measures are well understood so that the measures are taken seriously.

– To guard against an outcry by project participants on the possible use of KPIs for the purpose of punishing or congratulating people. Metrics should only measure the process and therefore should be anonymized (as far as possible). The improvements that will be taken by the analysis of these measures are intended to improve the work of the whole team and not of one part of the team at the expense of the other.

– To ensure that the burden of taking and compiling measurements is not too high so that this measurement does not impact the results of the measurements. It will therefore be necessary to measure only the smallest number of KPIs necessary, even if it means calculating the other KPIs.

8.10. Reporting attention points

Care should be taken to ensure that the usual tendency to blame the messenger for the content of the message is minimized. This involves considering several points of attention in the context of test reporting:

– the reporting audience, that is, for whom the reports produced are intended;

– how shall or could the reporting be used;

– impartiality;

– control of the evolutions of reporting throughout the project.

As you understand, reporting can provide information that does not satisfy project stakeholders. This may be the identification of a development inability to deliver the requested components or the expected corrections in a timely manner, this may be the identification of a large number of defects jeopardizing the delivery of components or systems which could be linked to late payment penalties, etc.

8.10.1. *Audience*

The execution of tests by testers, whether they pass or identify defects, does not add value if the results are not communicated to others. This can be to the project manager in order to inform them of the progress, or to the development teams in order to obtain the correction of the anomalies identified.

In a system-of-systems project, several hierarchical levels are interested in the test results, and each audience will have a different objective, which will imply different content depending on the audience:

– Development teams are interested in defects and releases to provide to test teams, as well as components that could benefit from a rewrite or refactoring.

– Test teams are interested in the execution of test activities (e.g. test case execution, test case re-execution, defect management) and also the design actions of the test cases. Tests, definition and requests of test data, etc.

– Operations and configuration management teams focus on the ability to deliver in the right environment, with the right data and on time.

– The business teams focus on the quality and relevance of the requirements and user stories, as well as on the quality of the elements that are/will be delivered to them.

– The various levels of management focus on:

- the ability of various teams to carry out the work;

- short-, medium- and long-term planning of the activities;

- elements that could affect them or their teams, positively or negatively.

8.10.2. *Usage*

Reporting should provide information that can help decision-making. As the tests continue to run after the measurements to create the reports have been taken, it is important to be factual and – if necessary – to mention the time of the image capture. Avoid, as much as possible, acrimonious exchanges based on the presentation of reporting information.

If reporting is used to punish or praise any teams, the teams providing the reporting data will provide you with data to avoid blame or pass it on to others.

It is therefore important to be impartial and unbiased in reporting, to limit yourself to facts, to measurable and measured elements. Obviously, you will use the measured elements to make forecasts, to see if the trends will allow – or not – to move on to the following phases while respecting deadlines and budgets.

It is often said that managers only understand diagrams with boxes and arrows. It is important to inform managers of the salient facts and their impacts, without drowning them in unnecessary details. The reporting must be adapted to the audience and the use, so the relevant information is not drowned in a large volume of less useful information.

8.10.3. *Impartiality*

Relations between stakeholders – co-contractors, subcontractors, etc. – are sometimes strained, and it is important that the information provided is accurate and understandable. There is a strong temptation to push the causes of the problems identified by one team onto other teams.

Political and relational aspects within the company and between the various co-contractors of a system-of-systems project are important. They are often associated, on the one hand, with a level of confidence, which implies that the data must be as impartial as possible but are also associated with the way in which the information is presented or communicated. We can assume that – by default – participants are all trying to do their best. Starting by blaming one or the other of the actors is not an effective solution.

It is therefore recommended to provide the data – the reporting – impartially and clearly. If a team seems to be the cause of malfunctions, I suggest notifying the team leader to inform them of this identification of malfunction and to collect the proposals for planned improvement actions. This will allow these improvements – positive actions – to be associated with this information – negative – to show that the teams are acting in a timely manner.

8.10.4. *Evolution of reporting*

Rome was not built in a day. Similarly, the implementation of test campaigns, and the related reporting, takes time and evolves as the campaign progresses (e.g. by adding new series of test cases).

There are two reporting areas to consider:

– definition of reporting for the project;

– definition of reporting for each iteration.

The definition of reporting at project level includes – for Agile projects – the definition of DOR (Definition of Ready) and DOD (Definition of Done), as well as the definition of the selected reporting modes (e.g. KANBAN board, burn down, burn up, velocity measurement, etc.). In sequential projects, it is important to define the periodic reporting that will be sent to stakeholders (e.g. progress in requirements coverage, achievement of milestones, use of resources, management of anomalies, etc.). It is at the beginning of the project that the objectives and targets, as well as the stakeholders, must be defined.

The reporting of each iteration – or each new version in a sequential project – can evolve over time, for example, by adding additional information.

8.10.5. *Scrum reporting*

This solution makes it possible to visualize the information necessary for each team, via KANBAN boards or other reporting methods. Scrum reporting is often seen only at the level of development progress, within Scrum teams, through wall charts. The fact that testing activities are activities to be included in the workload of the Scrum team (e.g. in the form of specific columns or tasks) implies the need for reporting of these activities.

In general, Scrum projects use burn-down type reporting (see Figure 8.9) and burn-up type reporting (see Figure 8.10).

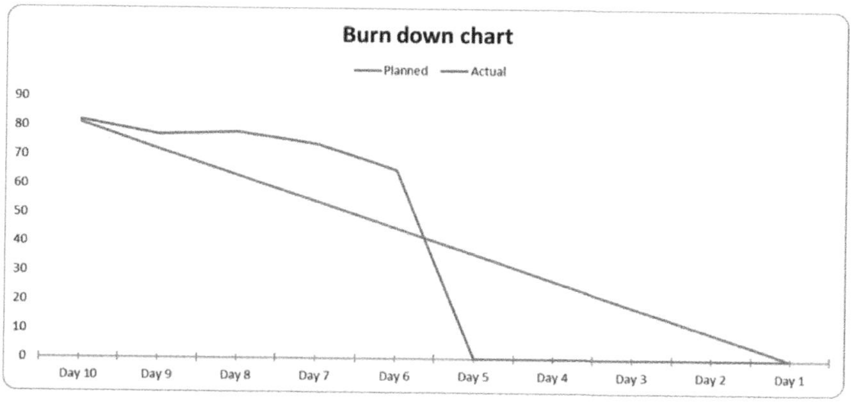

Figure 8.9. *Example of burn down. For a color version of this figure, see www.iste.co.uk/homès/systems1.zip*

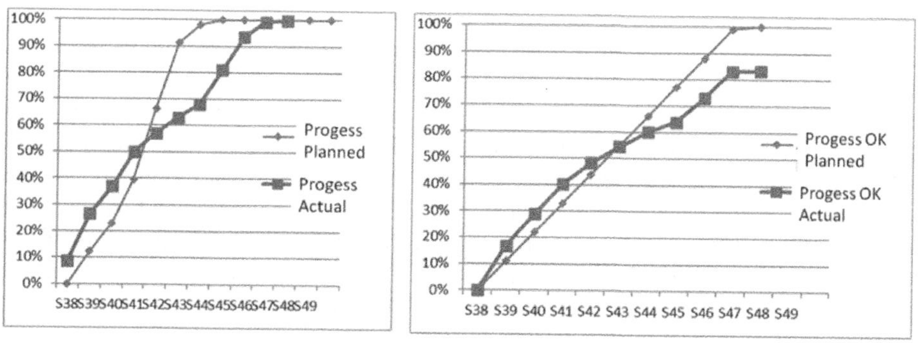

Figure 8.10. *Example of burn up. For a color version of this figure, see www.iste.co.uk/homès/systems1.zip*

8.10.6. *KANBAN reporting*

KANBAN reporting is mainly done – several times a day – via a "KANBAN board" where activities, tasks and their priorities are defined. The first column contains the entire backlog of objectives to be achieved in the form of Pos-tIt type cards, while all the other columns – except the last – represent the steps (or tasks) to be carried out to achieve the objective. The last column contains the objectives achieved. The vertical positioning in a column represents the prioritization of achievements within the same stage (the highest priority being at the top).

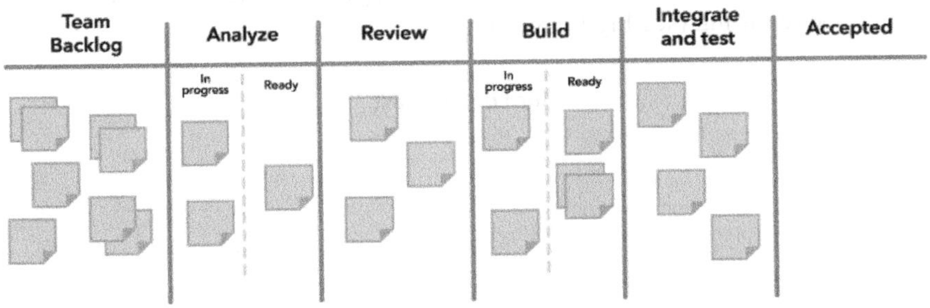

Figure 8.11. *Example of KANBAN board*

Moving from one column to the next – from left to right – represents the achievement of the column's exit criteria (column DOD); moving to the last column identifies that the objective has been achieved.

8.10.7. *Test design reporting*

Test cases must be designed before they are executed. It is therefore necessary to ensure that the design progresses correctly and is completed before the receipt of the components, applications or systems to be tested. The effort and time required for the design varies according to the level of quality of the inputs (e.g. presence and level of detail of the requirements), the complexity of the tests, environments and data as well as the need for automation. The level of detail will depend on the level of knowledge of the people in charge of performing the tests. The progress of the test design makes it possible to ensure that the tests will be ready at the start of the test execution phase and therefore anticipates this execution phase.

Among the test design aspects, it is necessary to ensure that:

– each of the quality characteristics – according to the chosen quality standard – has been covered correctly, both by positive tests (validating that the characteristic is present) and by exception tests (verification that exceptions are correctly handled);

– each test level is consistent and covers all necessary aspects, both functional and technical, anticipating performance and security aspects, as well as foreseeable evolutions;

– the nominal and degraded operations of each of the components, applications, subsystems or systems have been adequately covered, including combinations of malfunctions or degraded modes.

Reporting should include traceability from requirements to testing, to demonstrate requirements coverage, according to requirements' prioritization (e.g. criticality or functional importance) and component complexity. Reporting can be defined by component and/or by functionalities, in order to demonstrate a level of maturity of the elements.

8.10.8. *Test execution reporting*

Test campaign execution reporting must inform about the tests executed, differentiating:

– tests performed for the first time, to verify/validate components that have not yet been tested (tests);

– tests carried out to confirm the correction of components which previously generated one or more defect (retest);

– tests that could not be executed for one reason or another, for example, faulty previous test case, absence of prerequisites and/or dependencies, etc. (blocked tests);

– tests performed to ensure that components delivered in previous version continue to function in the same way (regression tests).

Execution reporting must also provide information on the workloads and times used to execute the campaign in question, as well as the differences between the planned workloads and deadlines and the actual workloads or deadlines. An identification of the level of coverage (including planned and actual) of the requirements is useful information for the various stakeholders.

8.10.9. *Reporting software defects*

Multiple types of reporting are required regarding defects:

– reporting on the detection and correction of defects over time, which makes it possible to identify if the detection and correction curves tend to meet or not;

– reporting on defects per software components, products, systems and subsystems, in order to identify possible defect clusters;

– reporting on the root causes of defects in order to identify the processes to be improved in order to avoid the future defect introduction;

– a report by component or functionality (sorting and search criteria) on the percentage of progress of the tests and the percentage of defects by measured criteria, in order to have a vision of the quality of these criteria;

– a report on the ability of the production teams to correct the backlog of anomalies within the deadlines set for the project and before the next deliveries.

Anomaly reporting is important to ensure that each anomaly is corrected, if possible, according to its level of impact on the project and on the product. The detailed analysis of anomalies, their introduction phases and their root causes make it possible to detect areas for improvement in the design processes. The analysis of the durations of correction and the increase in the backlog of anomalies make it possible to identify whether the delivery objectives will be able to be achieved.

In-depth defect analysis is useful for development teams to identify components and processes for improvement. Project managers can use this reporting to ensure that delivery teams are meeting their bug fix commitments (similar to SLAs). Test managers can use bug report analysis to adjust the level of information to put in bugs (details, environments, impacts, etc.).

For example, in Figure 8.12, we see that the trend line of the upper curve (sum of defects found per day) is divergent with respect to the trend line of the lower curve (sum of defects corrected per day). This involves detecting more defects than the development team can correct and deliver.

When associated with components, products, subsystems or systems, anomalies help identify those that are potentially of lower quality.

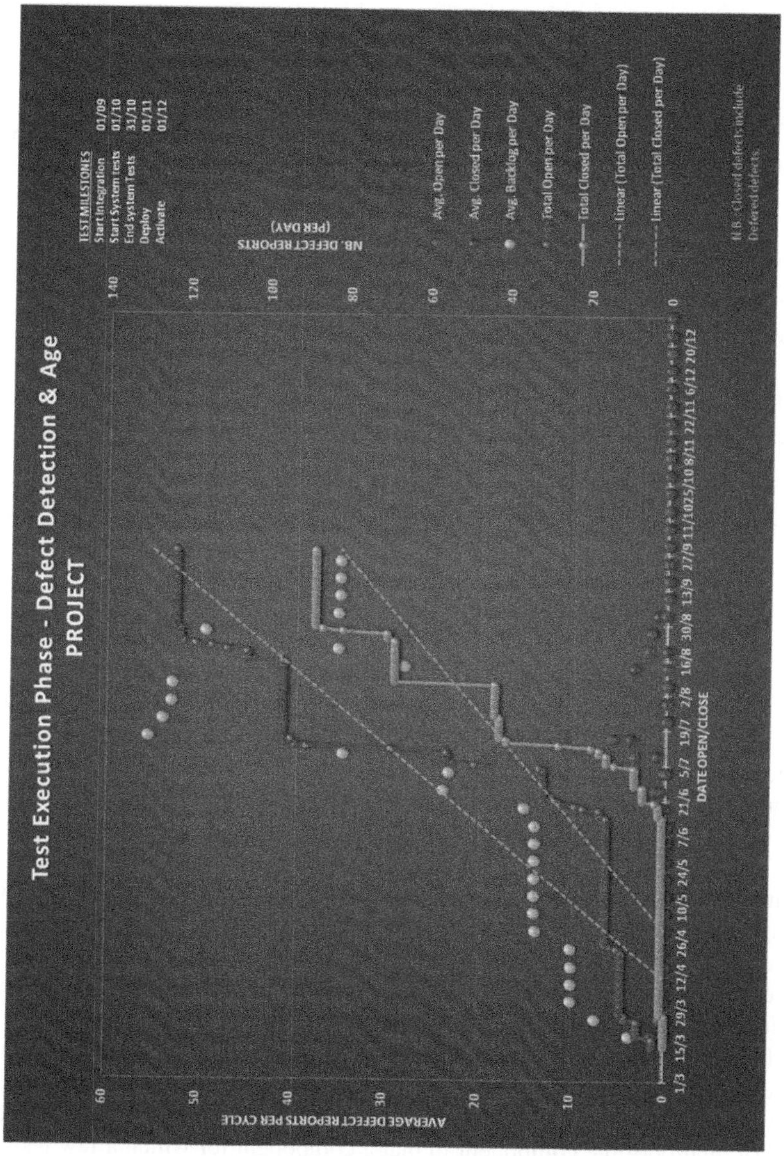

Figure 8.12. *Defects found versus fixed curves. For a color version of this figure, see www.iste.co.uk/homès/systems1.zip*

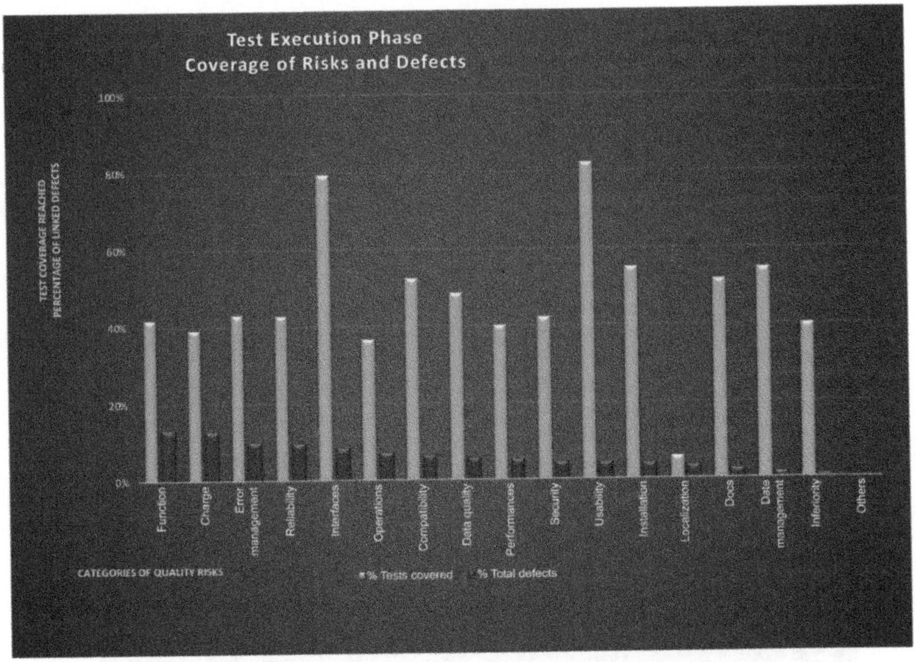

Figure 8.13. *Defects per component*

In Figure 8.13, we can identify that the number of defects is proportionally more important for the two components located on the left, compared to all the components. This could be due to a more extensive testing effort on these components and therefore detection of more defects. This is not the case; we can see it in Figure 8.14. This shows that it is necessary to use several graphs to correctly identify the causes of the problems.

Figure 8.14 is interesting in that it relates the percentage of anomalies to the percentage of progress of the tests, according to the components tested (the components can be replaced by the functionalities or typologies of anomalies). Each component is represented by a pair of bars. The sum of the red bars (right bars of each pair) represents 100% of the defects identified to date on the project. The gray bars (left bars of each pair) represent the percentage of tests performed on that component. What is important to compare are the orders of magnitude, the ratio between the gray bar and the red bar. It makes it possible to identify – indirectly – the level of quality of the component in that a component for which a few tests have been carried out and which has a high ratio of defects is probably not of very good quality (there is a risk of still finding many faults). On the contrary, a component where almost all of the tests have been carried out and where the percentage of defects is low can be considered of better quality.

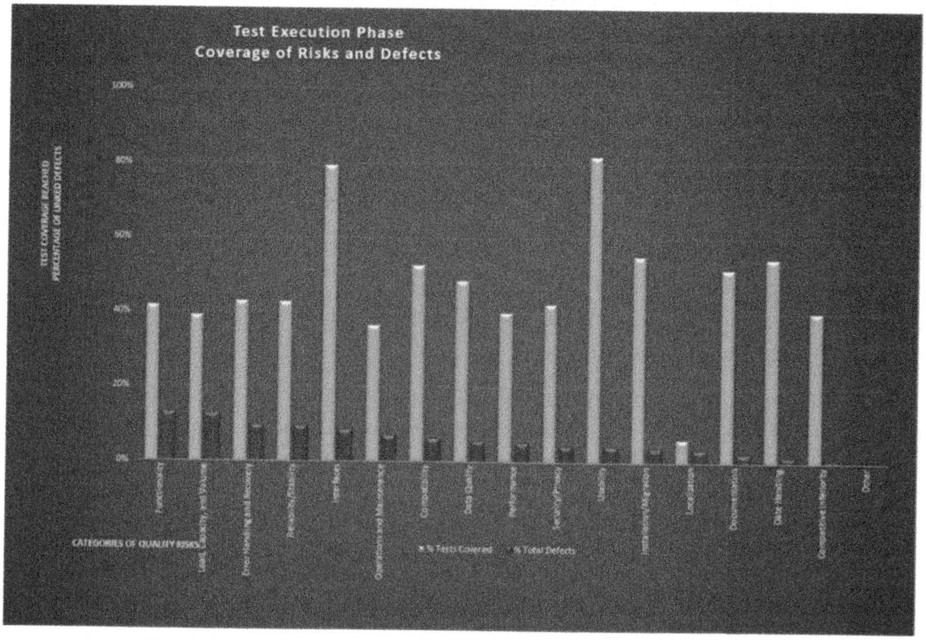

Figure 8.14. *Defects versus test coverage. For a color version of this figure, see www.iste.co.uk/homès/systems1.zip*

The comparisons performed on this type of graphs and these types of statistics are always performed "all other things being equal"; the criticalities and impacts of the defects are considered similar, and the sizes and complexities of tests and components are also considered similar.

Other graphs and statistics can be used from the anomalies, such as the identification of defects according to the phase (design, coding and code reviews or static analysis, unit tests, integration tests, system tests, integration of systems, acceptance tests), evaluation of the average age of defects (according to criticality) in order to identify compliance (or not with SLAs – Service Level Agreements), percentage of reopening of anomalies (identifying ping-pong effects between developers and testers or poor quality of bug reports), etc.

8.10.10. *UAT progress reporting*

User Acceptance Testing progress reporting is – as the name suggests – the responsibility of the users. As users do not have much time to devote to a task that covers a very large combination of test cases and test scenarios, there are various ways to perform these tests and report them. These ways can be combined:

– Users or their representatives can verify the test and quality proofs provided by the actors in charge of the design of the system or system-of-systems. This assumes that the documentation aspects are complete and exhaustive, with traceability from the requirements to the test proofs (on each version of software or component).

– The contractual aspects may include a period of pilot phase, or double work, during which the users in charge of acceptance use the system-of-systems and escalate the anomalies they identify.

– In the agile development models, a demonstration of the functionalities is planned at the end of each iteration. This solution is difficult to implement as part of a system-of-systems, given the need for hardware components.

Among the elements to be measured we have:

– the functionalities, requirements or User Stories validated, covered and validated in percentage and by period;

– the percentage of coverage of system-of-systems functionalities by end-to-end (E2E) tests by order of priority or criticality;

– the level of satisfaction of the operational, maintenance and support requirements of the system-of-systems, including in the context of deployment in extreme environments (extreme heat or cold, humidity, dust, etc.);

– information on anomalies and their corrections, compliance with SLAs by stakeholders.

8.10.11. *Reporting for stakeholders*

8.10.11.1. *Hierarchy*

To be effective, reporting to the hierarchy must focus on their needs in terms of monitoring and progress of the test project. This implies identification by the test manager of the necessary reporting axes. These axes often include:

– percentage progress in relation to the achievement of the final objective;

– identification of expenses already incurred and yet to be incurred;

– identification of risks and their trends of appearance;

– the level of quality of the input deliverables;

– process efficiency and improvement needs.

It is possible to use tools to facilitate the production of test progress reports for the hierarchy. In order to ensure that your reporting is independent of test automation tools and that it exactly meets the needs of stakeholders, it may be necessary to use spreadsheets that can integrate information from various monitoring tools.

8.10.11.2. *Development*

The points of attention of the development teams relate mainly to:

– level of quality of delivered components (e.g. by feature or sprint);

– number of defects identified, and their age (duration since their identification);

– backlog of defects to be fixed and the team's ability to correct them on time;

– number of times defects have been reopened;

– number of defects versus function points delivered, to identify if the teams are improving or not compared to industry and previous deliveries.

As with reporting to management, you can use tools to produce test progress reports for development teams.

8.10.11.3. *Users*

Users and customers focus primarily on:

– level of coverage and satisfaction of the requirements they have described;

– level of quality of the features they expect, depending on the priority level of these features;

– availability date for the features;

– available evidence to judge if the requirements have been met;

– costs and delays prior to implementation in production.

It is easy to realize that neither the hierarchy, the implementation teams, nor the customers care about the number of test cases executed by the testers. Therefore, reporting statistics based on test cases will be of no use for the hierarchy, the developers, or for the customers, especially since there is no linear relationship between the requirements and the number of test cases associated with it.

9

Testing Techniques

Before looking at the various test techniques, how can we group them, and what are their typologies?

9.1. Test typologies

I propose to synthesize the various types of test, each with a different utility and impact.

"Testing" should be seen as verification and validation ("V&V"). Verification and Validation both involve providing evidence that requirements have been met. This involves having measurable and achievable requirements. We speak of IV&V when the actions of V&V are carried out by an independent team. Validation focuses on an "intended use" which is not the case for verification. Tests can be dynamic (i.e. requiring the execution of the code of the application to be tested) or static (without executing the code to be tested).

Like any other activity, testing needs to be effective and efficient. Effective in terms of defect detection, efficient in limiting the effort required to achieve these objectives.

The development cycle does not matter on the tests: all the types of testing mentioned below can be carried out regardless of the development cycle. In a cycle (iteration, sprint, etc.), it is recommended to start with static tests and reviews, then to perform unit testing, then integration testing, then system testing, before doing application testing, systems integration, then system-of-systems testing and ending with acceptance testing. Each sprint after the first will have to ensure that the evolutions do not lead to regressions in what has been developed previously; this will involve (non)regression testing and sometimes automated testing. When implementing continuous delivery solutions (e.g. DevOps), it will be necessary to ensure, through systems integration tests and system-of-systems tests, that the end-to-end functionalities are not impacted. Performance, security and usability tests will also be necessary.

9.1.1. *Static tests and reviews*

Static tests and reviews are component verification activities that do not require that component to be running. These activities can use tools (we will talk about static analysis) or human beings (we will talk about reviews). The formalism and the level of detail may vary, leading to a variation in the effectiveness of these activities.

– Use: effectiveness of reviews has been demonstrated since the 1970s (search for Michael Fagan) and ranges from 50 to 85% defect detection. These activities can be carried out even before the software is executable, therefore very early in the delivery, which leaves more time for corrections.

– Impact: early defect detection, including design defects, saves a lot of time and increases efficiency.

9.1.2. *Technical tests*

Security, usability, performance testing, pentest, etc. can exist at each test level and for each system-of-systems level of granularity, and focus on – non-functional – characteristics described in standards such as ISO25010.

– Use: all the quality characteristics have a greater or lesser interest in the short, medium or long term for the stakeholders. The level of importance and therefore the investment one should make will vary depending on the environment.

– Impact: as with functional testing, these activities should be performed early in the development cycle, to provide information as early as possible and allow more time to implement fixes.

– Positioning: some technical tests cannot easily be implemented early in the development cycle. Pentests, for example, can hardly be carried out if all the components are not developed and deployed in an environment. Nevertheless, the rules verified by the pentest can be verified by static tests and/or reviews. Similarly, performance tests are often handled at the end of development when early performance checks are possible (e.g. budgeting of component response times).

A comparative analysis of the usefulness, impact, effectiveness and efficiency of each type of test would deserve an entire chapter. If we want to simplify, we could compare the types of tests to increasingly fine-mesh filters. Any deletion of a filter (any deletion of a test type) will increase the number of defects to be identified by subsequent filters, otherwise they will be identified by end-users. We can therefore see that several types of tests and several levels of testing are necessary to ensure the quality of the software delivered.

9.2. Test techniques

Many testing techniques exist, some dating back more than 40 years; Glenford Myers' book *The Art of Software Testing* (Myers 1979) is still useful and applicable, for example:

– equivalence partitioning (EP);

– boundary value analysis (BVA);

– decision tables (DT);

– states and transitions (graphs and tables) (STT).

These testing techniques have also been detailed in my book (Homès 2012).

Other fundamental techniques exist and will be detailed as follows:

– CRUD;

– Paths.

Several other test techniques exist and are detailed in works such as Koomen et al. (2006) and Koomen (2008):

– decision testing, condition testing, code, loops, components and calls, messages and exceptions, etc.

In general, test techniques can be combined with each other. As the objective is to find defects as early as possible (also called "Shift Left"), it is recommended to select the most suitable techniques – the most effective and the most efficient – for the level of test.

Each test technique focuses on one or more types of defects. Several testing techniques should be used to identify as many types of defects as possible. Whatever techniques are used, it is mandatory to ensure that these techniques are correctly implemented. This involves identifying the number of tests that can be performed with each technique and the number of tests performed using that technique. We will then talk about the coverage of the test technique.

9.3. CRUD

The CRUD acronym applies to the testing of data and their life cycle. It means Create (create), Read (read), Update (update) and Delete (delete), which represent the order of data management.

To ensure the reusability of test data, it must be ensured that the data created – at the start of a test or of a campaign – is deleted at the end of the test (or of the campaign) in order to allow a new execution with the same data.

Measuring this technique includes measuring the number of existing data types and their coverage for each C-R-U-D operation.

In systems-of-systems, data is often created in one system, provided through interfaces and updated by other systems, and – optionally – deleted by those systems. In this case, it will be necessary to create a matrix to identify, for each data, the systems that can impact it, and for each component or system, the data used or manipulated. These two matrices will make it possible to identify the components, systems or products to be retested in the event of a modification in the chain using one or more data series.

9.4. Paths (PATH)

Path-based techniques can be seen as the application to functional testing of the white-box techniques "Instruction Coverage" and "Branch Coverage". You can apply the path technique to functional testing (i.e. black box tests) at any test level, including end-to-end testing (E2E testing).

9.4.1. *Operation*

To represent the various paths and interactions between systems or subsystems, we can use either a swimlane representation as shown in Figure 9.1 or flow diagrams (Figure 9.2).

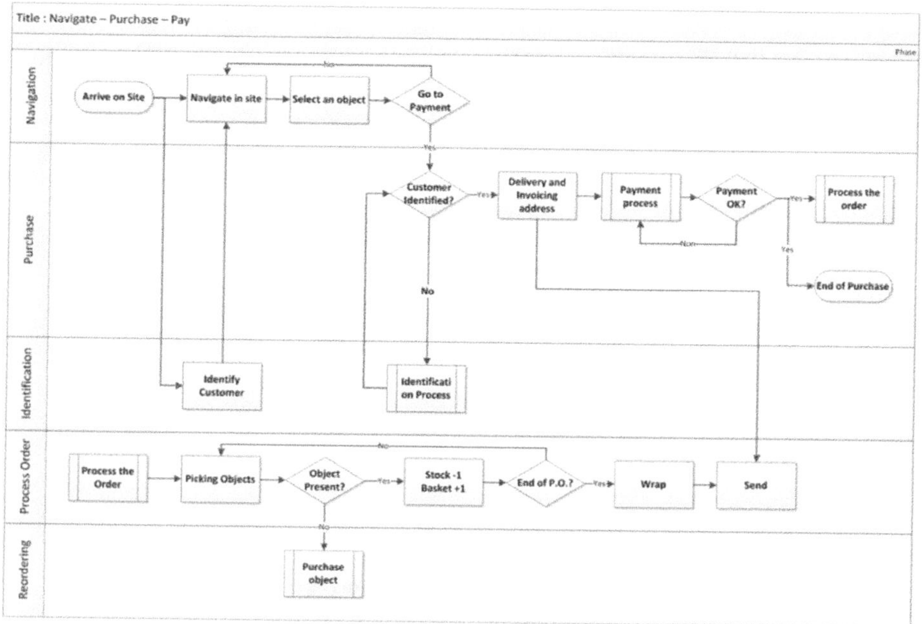

Figure 9.1. *Swimlanes*

9.4.1.1. *Functional paths (black box)*

Instead of focusing on the instructions, loops or functions in the code, we will focus on the functionalities of the component, product or system (with the messages, screens etc.), which are testable in the environment.

The principle is to go through all the paths of the application by focusing on the most probable sequences (called "Happy Path") and the paths leading to exceptions.

Example: the standard path (happy path) of a purchase on an e-commerce site is:

1) search the catalog;

2) select one or more item(s) and put them in the basket;

3) carry out the invoicing operation, detailing delivery addresses, invoicing, etc. as well as payment information;

4) complete the payment transaction successfully.

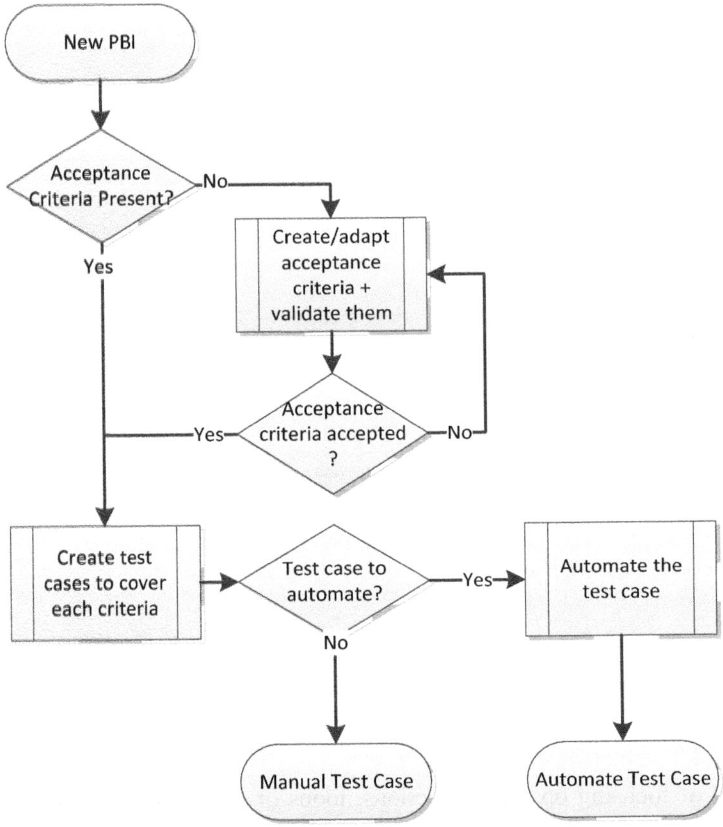

Figure 9.2. *Flow diagram*

Exception or invalid paths may include:

– canceling the transaction at any time;

– deletion of items already selected in the basket;

– non-receipt of messages;

– introduction of invalid addresses or payment methods;

– refusal of payment validation (when two-level authentication is implemented);

– timeout at any time during the transaction.

The combination of valid paths is immense in a system-of-systems. It is necessary to define the tests to be automated (e.g. to create a set of tests to be executed with each new version delivered), and to define the subset of functionalities that must necessarily work (priority paths).

The advantage of this technique is that it is mainly based on the user interfaces and makes it possible to check the operation of the entire system, including the back office. The disadvantage mainly relates to the number of paths to be considered since this technique should be combined with the techniques of equivalence partitions (EPs) and BVA then with techniques such as decision testing tables (DTTs) and the calculation of the cyclomatic number of functional paths that can be considered on the system – or subsystem – to be tested.

The coverage measure of the paths technique is to note the number of paths executed divided by the number of possible paths. We can detail according to the priority of the paths and according to what the paths contain (in the white-box test).

A graphical representation of the functional paths makes it possible to design a directed graph where each node is a set of actions (without branching) and each branch is an arrow connecting the nodes between them. This representation can be made from process analysis or from swimlane-type representations (swimming lines). The analysis of the graph will identify the cyclomatic number [V(G)] associated with the graph, and thus determine the minimum number of test cases necessary for coverage of the entire graph. The analysis of the decisions, and of the conditions that make them up, can be synthesized in a DTT which will make it possible to determine, for each relevant variable, the data to be considered (by combining the techniques of equivalence partitioning, limit values and domain testing). The combination of the various techniques will make it possible to define the minimum subset of tests to be designed to cover each of the functional paths of the system and to specify the types of data (scores of equivalences and limit values) which are necessary to pass through each of these paths. It is obvious that on a system-of-systems, the number of combinations will – probably – be too large and that we will have to concentrate on each of the subsystems.

It is therefore possible, by combining basic test techniques, to offer a measurable level of end-to-end (E2E) test coverage.

9.4.1.2. *Technical paths (white box)*

"White-box" techniques: coverage of instructions, decisions or branches and conditions, have already been dealt with in Homès (2012) and in Homès (2021). They mainly apply to unit tests. As part of systems-of-systems testing, they should provide evidence of correct execution of unit tests. These tests are important because they allow defects to be discovered as early as possible (shift left).

These techniques for covering paths, decisions and branches apply to directed graphs. These same directed graphs – and the techniques mentioned above – can be applied to functional aspects.

9.4.2. *Coverage*

Achieving full path coverage is usually not possible, so it will be necessary to define:

– the nominal path(s); and

– one or multiple exception paths.

The principle is to cover each decision in true and false, from the beginning to the end of the process. It is recommended to use the other test techniques described below and in Homès (2012) at the same time to cover as many combinations as possible in a small number of test cases.

The measure of achieved coverage is calculated as follows:

$$Coverage\ reached = \frac{Paths\ covered\ by\ at\ least\ one\ Test\ Case}{Combination\ of\ all\ possible\ paths}$$

The objective is to be as close as possible to 100% coverage.

9.4.3. *Limitations and risks*

In the context of end-to-end tests (E2E tests), the combination of possible Paths – especially with the combination of data to be used – makes complete coverage difficult to achieve. In critical security areas, it will be necessary to achieve the level of coverage required by the applicable standards according to the level of criticality of the system.

Often, it will be necessary to use the risks to identify the order of execution of the path tests, to ensure the most consistent coverage possible.

9.5. Equivalence partitions (EP)

The EP test technique is simple to understand, but more difficult to apply correctly. The principle is, for each variable, to group together within the same partition all the values which will be treated (processed) the same way. The goal is – as with all testing techniques – to limit to a reasonable (finite) number the infinite number of tests that one can run, for example, by testing every possible value for every variable. Note that equivalence class is synonymous to EP.

Let us take as example the formula: $y = \sqrt{1/x}$ that can be described as the following code: `y=sqrt(1/x)`.

We have:

– three EPs for the value of x:

- one valid partition with x positive, which will give a valid Y result,

- one invalid partition with negative values for x, which should give an error message (imaginary square root),

- one invalid partition (singleton) where x has zero value, which should give an error message (zero divide);

– two EPs for the content of variable x:

- one valid partition with x numerical,

- one invalid partition with any content of x other than numerical;

– three EPs related to the size of the data contained in x variable:

- one invalid partition for sizes below the minimum size for x,

- one invalid partition for sizes above the maximal size for x,

- one valid partition for sizes of x content between minimal and maximal size;

– two partitions for the values of y:

- one partition for the acceptable numerical values of y variable,

- one partition for the invalid (i.e. non-numeric) values of y to be rejected.

We could think of other ways to group the values in partitions.

You might think that the size aspects of a variable's contents are trivial, but the behavior may have to be taken into account depending on the use of the variable (e.g. minimum and maximum size of a password) or depending on the definition of the variable (e.g. what is the behavior of the data entry field if the capacity of this variable is exceeded? Will there be an overflow resulting in a security failure? These types of failures are a gateway for hackers).

Consider another example, validating a date as DDMMYYYY:

For the days, we have:

– a valid partition with days: 01, 02, 03, …, 28 because all months have at least 28 days;

– a valid partition for day 29;

– a valid partition for day 30;

– a valid partition for day 31;

– an invalid partition for any value of DD below 01;

– an invalid partition for any value of DD above 31;

– an invalid partition for any value of DD where DD is decimal;

– an invalid partition for any value of DD that is not numerical.

This implies that numeric values preceded or followed by a space are not accepted.

For the months, we have:

– a valid partition for months 01, 03, 05, 07, 08, 10 and 12 that have 31 days;

– a valid partition for months 04, 06, 09 and 11 that have 30 days;

– a valid partition for month 02 that may have 28 or 29 days;

– an invalid partition for any month value below 01;

– an invalid partition for any month value above 12;

– an invalid partition for any month value below that would be decimal;

– an invalid partition for any month value that is not numerical, thus numerical values preceded or followed by a space are not accepted and will be included in this partition.

For the year value, we have:

– a valid partition for any positive or negative year value;

– an invalid partition for year zero;

– an invalid partition for any year value that is not numerical, or preceded or followed by non-numerical data;

– an invalid partition for decimal year (with a value after decimal point).

9.5.1. *Objective*

The main purpose of testing EPs is to find partitions that would not be handled correctly by the component, product or system. As the principle is that all the values of an EP are treated in the same way, by taking only one value from each of the partitions (a representative of the partition) we will limit the number of test cases to be considered, while covering all partitions.

NOTE.– The examples have focused on numerical values, but the EPs can also relatc to temperatures, weights, time, etc.

9.5.2. *Operation*

The principle is to cover each of the valid and invalid partitions at least once. We will therefore combine the values for DD, MM and YYYY, by combining them with each other. Care should be taken, when including an invalid partition, to combine it only with valid partitions. This will ensure that the invalid partition is actually showing the expected error message.

It is possible to combine several invalid partitions together, but this increases the combinations to consider. This can be the subject of exploratory tests.

9.5.3. *Coverage*

Coverage is calculated by measuring the number of partitions covered divided by the total number of valid and invalid partitions.

9.5.4. *Limitations and risks*

It is obvious that, given the very large number of variables used in a component, the coverage of EPs is recommended for the level of unit tests. However, it is possible to consider partitions in functional areas when testing for other levels.

The main risk with this technique comes from a bad definition of the partitions which can lead to the choice of an incorrect representative.

9.6. Boundary value analysis (BVA)

BVA is very efficient and finds faults at the boundaries of valid ordered EPs. This technique therefore requires the prior implementation of the technique of EPs.

9.6.1. *Objective*

Check the software behavior at the boundary of valid ordered EPs, with the following three values:

– the actual boundary value;

– a value corresponding to the boundary minus the smallest increment possible;

– a value corresponding to the boundary plus the smallest increment possible.

9.6.2. *Operation*

Identify, for each ordered EP, the boundary values – lower and upper – and the size of the increment.

Identify the expected behavior at each limit, above and below each of these limits.

Design a test for each of these values. In some cases, the tester will have to manipulate the content of the variables to ensure that each of the values is tested.

9.6.3. *Coverage*

Cover at least once each of the three values. This will require identifying, for each variable of the component or product, or even of the system and its external interfaces, the partitions of equivalences to be considered and the limits to be covered.

9.6.4. *Limitations and risks*

BVA should be implemented after equivalence partitioning.

Unordered EPs have no boundaries, so this technique is not applicable to them. Lists of values (e.g. in drop-down lists or "Case" statements) also have no limit values.

9.7. Decision table testing (DTT)

DTT is a powerful technique to test combinations of functional decisions – business rules – in detail. The repository used to design the tests (the test basis) contains the conditions and decisions to be processed. The purpose of the decision tables is to identify the combinations of conditions and decisions that will need to be covered by tests. The following checklist can be used to ensure that decision tables are specified correctly:

– Are the conditions influencing the process clearly recognizable and have they been described?

– Is the output of the process clear and is it possible to predict the results?

The decision table testing technique can be used at any level of testing. It can generate a large number of combinations to be tested, so it may be useful to simplify the resulting tables in order to keep a reasonable number of combinations.

9.7.1. *Objective*

The objective of the decision table test is to cover the combinations of business rules with the smallest possible number of test cases, without forgetting conditions.

9.7.2. *Operation*

For each complex business rule using two or more variables, identify valid and invalid EPs, combine these partitions into a table, and simplify the table by grouping columns with the same conditions (except one) and the same results or actions.

9.7.2.1. *Example*

Imagine a decision table that determines when an individual can drive a four-wheeled motor vehicle in France. We will only take into account what relates to the driving license. The management rules relate to (1) the age of the driver (prohibited if under 15 years old, authorized if accompanied by an authorized companion between 15 and 18 years old, no restriction if 18 years old and over); (2) on successfully passing the theoretical exam known as the "code"; (3) the presence alongside the individual of an authorized companion (e.g. a parent for accompanied driving or a driving instructor) and (4) successfully passing the practical exam, which can only take place at the age of 18. The results can be Yes, driving is allowed or No, driving is not allowed.

We can therefore describe the table in Figure 9.3 with each line covering the four management rules and relating to 24 columns, that is, the product of the number of partitions for each variable 3 x 2 x 2 x 2.

	01	02	03	04	05	06	07	08	09	10	11	12	13	14	15	16	17	18	19	20	21	22	23	24
Age A: <15 15<B<18 C: 18	A	A	A	A	A	A	A	A	B	B	B	B	B	B	B	B	C	C	C	C	C	C	C	C
Theory OK ?	Y	Y	Y	Y	N	N	N	N	Y	Y	Y	Y	N	N	N	N	Y	Y	Y	Y	N	N	N	N
Accompagn. present?	Y	Y	N	N	Y	Y	N	N	Y	Y	N	N	Y	Y	N	N	Y	Y	N	N	Y	Y	N	N
Licence obtained?	Y	N	Y	N	Y	N	Y	N	Y	N	Y	N	Y	N	Y	N	Y	N	Y	N	Y	N	Y	N
Driving Allowed?	N (a)	N	N	N	N	N (a)	N (a)	N (a)	N (b)	Y	N (b)	N	N (b)	N (d)	N (b)	N (d)	Y	Y	Y	N	N (c)	N (d)	N (c)	N (d)

Figure 9.3. *Expanded decision table*

The decision table can be simplified as follows:

a) Whatever the values of the scores (2), (3) or (4), rule (1) defines that if the individual is under 15 years old, they cannot drive. This merges columns 1 through 8 into one.

b) Regardless of the age of the driver, if they have not passed the theory test, they are not authorized to drive. This allows us to merge columns 13, 14, 15, 16 and 21, 22, 23 and 24 into one.

c) If the driver is under 18, they cannot have a driving license, so columns 9 and 11 can be merged into one.

d) Regardless of the age of the driver who have passed their theory test, if they have not passed the practical test and is not accompanied by an authorized companion, they may not drive. This merges columns 12 and 20.

This limits the number of test cases to be executed from 24 to 8 as shown in Figure 9.4. This reduces the test load.

Some complex management rules include more than four variables. Decision tables make it possible to identify inconsistencies in management rules that combine: As part of a merger of aeronautical standards relating to flight bans, we have implemented a decision table comprising 212 columns. The detailed analysis of this table identified three cases where, despite compliance with all the standards, the flight restrictions were not respected.

	01	09	10	12	13	17	18	19
Age A: <15 15<B<18 C: 18	A	B	B	B	B	C	C	C
Theory OK?	Y	Y	Y	Y	N	Y	Y	Y
Accompagn. present?	Y	Y	Y	N	Y	Y	Y	N
Licence obtained?	Y	Y	N	N	Y	Y	N	Y
Driving allowed?	N (a)	N (c)	Y	N (d)	N (b)	Y	Y	Y

Figure 9.4. *Simplified decision table*

9.7.3. *Coverage*

Decision table coverage is achieved by running one test per column of the simplified table. We therefore measure the number of test cases executed compared to the number of test cases planned.

9.7.4. *Limitations and risks*

The test based on decision tables is based on a good identification of the management rules, the variables impacted and the results to be obtained. This implies a good identification of the EPs (even of the limit values if they impact the management rules) and a rigor in the simplification of the columns: indeed, it is necessary to avoid deleting a column which should not be. For this, we can give each column of the initial table a weight of 1 and add the weights of the merged columns. For Figure 9.4, the weights of 8 for column 01 (it covers eight test cases in total), a weight of 8 for column 13, a weight of 2 for column 09 and a weight of 2 for column 12. The other columns having a weight of 1, the sum of the weights gives 24 which corresponds to the number of columns of the initial table.

When running tests, it is recommended to start with the columns with the highest weights as they cover more test cases.

You will find more explanations in Homès (2012).

9.8. Use case testing (UCT)

UCT is often used to describe how a User Story is expected to work (in agile mode) or how a user should use the component, product or system. They are often defined from the acceptance criteria of the User Story. They also represent – in a sequential mode of development – a description of how to use the system or system-of-systems.

9.8.1. *Objective*

Verify the operation of components, products, systems or systems-of-systems as they might be used. This involves understanding the objectives, constraints and habits of use and mastering the direct flows – generated by the test – and indirect – generated in parallel to the test case by other users or systems. It will therefore be necessary to break down the system to understand its elements, functionalities, connections and exchanged flows. Then, it will be possible to design use cases covering the elements, systems and flows in a realistic way.

9.8.2. *Operation*

From the description of the use case and the decisions that can be applied, define a graph of operation and cover the graph with the necessary paths. Ensure that all detail elements are present to determine that the use case is suitable. This may involve using a standard template for the project or organization. The following are some points that can serve as a basis for a checklist:

☐ Is the template needed for the project or organization filled out completely?

☐ Is the use case diagram (graph) available?

☐ Is the use case a separate task in itself?

☐ Is the purpose of the unit test clear?

☐ Are the actors involved for this use case identified?

☐ Is the use case tied to the functionality and not a sequence of screens?

☐ Have all known exceptions been described?

☐ Does the use case include a detailed step-by-step plan?

☐ Has each stage of the scenario(s) been described clearly, completely and unambiguously?

☐ Are all the actors and all the steps, mentioned in the use case, necessary for the execution of the task?

☐ Are all the steps executable?

☐ Are the results of each step verifiable?

☐ Do the pre- and post-conditions match the use case?

9.8.3. *Coverage*

Calculate the number of paths – valid and exception – for each use case, and ensure that each of these paths is covered by at least one test.

Divide the number of paths covered by the total number of possible paths.

9.8.4. *Limitations and risks*

Often the use cases are not correctly and filled out. This leads to ambiguities and disagreements between test and development teams. Use cases must be validated by the business representatives (PO or PPO) as well as by the technical architects before being provided to the development and test teams.

Defining a list of use case acceptance criteria is one way to do it, but can be considered a hindrance to development.

9.9. Data combination testing (DCOT)

9.9.1. *Objective*

The DCOT technique can be used both at a detailed level (e.g. at the level of component testing or integration testing) and at the system-of-systems test level. All types of system feature information can be used. It focuses on combinations of data in variables.

9.9.2. *Operation*

From the formal system documentation (e.g. functional design, logical data model, requirements), informal system documentation (e.g. manuals, memos, emails, etc.) and/or domain expertise (frequently undocumented), design the nominal data combinations and the combinations that can generate exceptions. It is recommended to use the architecture diagram – the processing and data flows – as well as the data schemas, in order to design use cases covering all data combinations.

To limit the combinatorial explosion, you can use combinatorial limiting techniques (combinations of pairs, orthogonal arrays, etc.).

9.9.3. *Coverage*

Measure the combinations covered by test cases against the total combinations to be covered.

9.9.4. *Challenge*

The design of the data sets is the main challenge here. In the context of systems-of-systems operating in real time with the IoT, the generation of test data consistent with each other is a major challenge.

9.10. Data life cycle testing (DCYT)

9.10.1. *Objective*

The main objective of DCYT is to find faults in the use and integration of data within the various systems and subsystems. This technique is useful in the search for functional defects, and also connectivity and usability. DCYT is very effective after functional testing of components, to find integration defects. The most important test basis is the CRUD matrix and the description of data integration rules. In the absence of the matrix, it must be created to allow the design of the tests which are the basis of DCYT.

9.10.2. *Operation*

Identify test situations and cover them with CRUD type tests. Then, determine the data usage (their creation, reading, updating and deletion) in the various functions and then test each of the test situations separately in each of the functions.

The test cases should include checking the data integrity rules, and make sure to cover each of the decisions. It will then be necessary to determine the logical test cases and ensure that each of the data passes through each of the states. The next phase will be to create physical test cases and define a common starting point.

9.10.3. *Coverage*

For each data field (variable) defined, ensure that each of the states (CRUD) and each function, functionality or decision using this data has been correctly covered by at least one test. For each variable, ensure that the transition sequence between each of the states of the data life cycle is consistent.

9.10.4. *Challenge*

Data cycle evolution usually happens for sets of data, not just for data in isolation. Some data may evolve in one or more data ranges (data in a normal range, and around the normal range, or even beyond these normal ranges). In systems-of-systems, it is also necessary to measure the functioning if the data are not available, which can show a failure in the networks or in the sensors.

9.11. Exploratory testing (ET)

ET has been proposed by Dustin (2003) and Whittaker (2009) and others.

You seldom have an exhaustive and detailed description of the requirements, their relationships and dependencies. The tester is often required to use their knowledge of the business domain, their intuition, their knowledge of test techniques and types of attacks, in addition to their analytical skills, to identify the specificities of the product. This involves exploring the product to understand how it works and gaining knowledge that will be used to design effective test cases.

ET is not really a testing technique, as it is not based on an existing baseline technique. This absence of a formal reference system leaves the choice to the tester on the basic test techniques to be applied, and therefore does not provide any guaranteed level of measurable coverage. Often, ET is associated with testing performed in the absence of a formal test repository (requirements, specifications, user stories, etc.) such as detailed design documents. However, this is not always the case: the functionality is explored before starting the tests, so the tester will only have as a reference what has been developed. It will therefore be impossible for them to identify features whose development has been forgotten, or for which the actuators (controls, buttons, etc.) are missing. The analysis – the exploration – of the system will be based, in addition to the existing documentation if it exists, on the existing system.

ET identifies test conditions iteratively, using the results of previous tests to guide subsequent explorations and tests. The testers will use all the testing techniques at their disposal to select, during their exploration, the most appropriate techniques.

9.11.1. *Objective*

Experienced testers are frequently able to find defects by exploring applications, components, products or systems. Tests performed are based on intuition, experience and exploration of the items being tested.

9.11.2. *Operation*

Exploratory tests can use “attacks” as defined by Whittaker (2003) and Hagar (2013) and also all the techniques present in this chapter, or even others. A frequent element during ET is the concept of test sessions: time slots – from 30 to 90 minutes

– allocated to the testing of a part of the component, product or system, focused on one or more objectives defined in a "charter".

This technique allows the test manager to define coverage based on the risks associated with each feature of the component, product or system. The higher the risk, the greater the duration and number of sessions.

9.11.3. *Coverage*

It is not possible to measure the coverage of exploratory tests in terms of the techniques to be used because the techniques used by testers vary. However, it is possible to measure the coverage of the charters, by comparing the charters carried out with the charters planned.

If the coverage of functionalities by charters is complete and consistent (e.g. each functionality is covered according to its criticality by one or more charters, absence of gaps or groupings of charters), then the analysis of the coverage of charters (charters covered on total charts) can provide an estimate of product quality.

9.11.4. *Limitations and risks*

Exploratory tests are mainly carried out during the "systems" test phase, that is:

– when the software components have been properly integrated with each other;

– when the application can run on the tester environment;

– late in the development cycle.

The objects tested will be the software component(s) operating in a test environment as representative as possible of production.

Reference documentation is often missing and defined by the tester based on their own experience, for example: behaviors consistent with or similar to competing products, as well as all the documents available and/or referenced on the project.

– All basic and advanced testing techniques can be used as needed by the tester.

Limitations: like any test technique requiring the execution of the component to be tested, this component must be present and be able to run in the test environment, with test data and correct configuration.

– It is impossible to define a level of coverage for exploratory tests, to plan their duration or scope based solely on the experience or intuition of the testers. A workaround is to define test sessions of a limited duration with clear test objectives associated with a functional area to be tested, in order to cover all areas fairly. When planning tests, the test manager will be able to define the effort to be made according to the level of risk of the functionality to be covered. During the execution of the tests, the tester will have to limit themselves to the planned duration.

– As the intuition and experience of the tester is put to use during the execution of the test, it is recommended to use testers who are experienced in both the testing techniques and the functional domain.

Advantages: the exploratory test is very pleasant to perform because it allows each tester to show their own skills, experience, intuition and ability to identify many defects quickly. In addition, the exploratory test can identify defects that were not anticipated.

Several books are available to develop your understanding of ET: Dustin (2003), Whittaker (2003, 2009) and Hendrickson (2013).

9.12. State transition testing (STT)

The test of states and transitions between states (STT – State Transition Testing) makes it possible to model the behavior of the system by passing – through transitions – from a starting state to a final state. A graphical representation of states and transitions makes it possible to share between the production and test teams on the one side and the business and PO teams on the other side. A representation in a table of states and transitions makes it possible to identify missing transitions.

STT is effective for testing finite state systems and most complex applications or systems. This technique also makes it possible to identify missing transitions between various states.

You will find more information on this technique in Homès (2012).

9.12.1. *Objective*

Identify the smallest number of paths that will cover each of the states and each of the possible transitions of the application, product, subsystem or system. Run the test for these paths.

9.12.2. *Operation*

Identify each of the states, including transient states.

Identify each of the possible transitions for each state and the criteria that trigger them (test data, states, etc.).

Create an STT table to identify missing transitions by listing, for each system state, all possible system transactions, and the resulting states (and actions).

Design as many paths as necessary to cover each transition and each state.

9.12.3. *Coverage*

The coverage is calculated by measuring the number of paths covered divided by the sum of the paths covering each of the transitions and going through each of the states.

Be careful" combinatorics quickly becomes very important in systems and justifies the automation of these tests. On the contrary, if the user interfaces are physical interfaces (buttons, levers, etc.), automation may become impossible or extremely complex (e.g. requiring the creation of mechanical simulators, robots or specific actuators).

9.13. Process cycle testing (PCT)

PCTs are implemented to test quality characteristics such as relevance, that is, the integration between the information system and the administrative organization that will use it.

9.13.1. *Objective*

Ensure that the administrative actions that the information system must facilitate are carried out correctly, produce the expected result and that the various functional paths – nominal and exceptional – can be carried out.

9.13.2. *Operation*

The main technique used here is the path technique. The recommended depth level is two, but you can go deeper or deeper. Processes can be described in swimlanes (swimming lines) and each action or step in the process is identified, with decisions, manual or automated processes and supports (documents, etc.).

We will:

– identify test situations;

– create logical test cases;

– create physical test cases;

– establish the starting point;

– run the tests;

– check that the paths have been followed correctly.

9.13.3. *Coverage*

All valid paths have successfully been traversed.

All invalid paths have correctly identified as impossible.

9.13.4. *Limitations and risks*

The following checklist can be used to ensure that the PCT technique is correctly implemented:

– All manual procedures to be performed by users have been described?

– Has the responsibility and authority for each manual task been described?

– Has each individual task been described?

– Have the safety aspects of this procedure been described?

– Is the triggering element clearly identified (i.e. the one that will trigger the procedure)?

– Have the data and forms to be used, as well as their sources, been identified?

– Have the activities to be implemented been described, including exceptions and checks?

– Is the result of each procedure clear?

– Have the various decision points been described, including the conditions attached to them?

– Is a distinction made between user procedures and system procedures?

– Have the relationships between automated and non-automated parts of the information system been identified?

– Is a user manual available (even in draft form)?

– Are degraded processes envisaged?

9.14. Real life testing (RLT)

9.14.1. *Objective*

The purpose of RLTs is to simulate statistically realistic use of the system in a responsible manner. This is not to simulate the behavior of the system in separate or independent situations, but to simulate a realistic use of the system. These tests focus on non-functional characteristics such as efficiency, connectivity, continuity and performance of the system under test. It is necessary to define the usage objectives – the usage profiles – according to the capacities of the production systems in order to test whether the system is capable of managing realistic volumes. To find with this type of tests, we have:

– transaction failures after use for a long time;

– failure of transactions if they are executed in a given sequence;

– too slow response times or insufficient execution speed;

– lack of memory or insufficient capacity of storage space;

– concerns of bandwidth and communication network.

9.14.2. *Operation*

The RLT is part of the acceptance tests. In the world of aeronautics, it is ground and flight tests, while in the naval world, it is the sea trial campaign.

In general, you will need to define realistic and predefined load profiles. These profiles are often split into operational profiles and load profiles. The basic technique here is to simulate a statistically realistic use. You will therefore have to:

– identify groups of users, including system administrators and operators in charge of backups;

– associate types of activities with user groups;

– determine the frequency of execution of activities per period, which can be determined by statistical analyses on previous versions of the system;

– determine the relative frequency to have consistent distributions between the various types of activities.

From the above information you will design physical test cases from user scenarios (or business scenarios), using data (files, transactions, etc.) of a realistic volume compared to the production, as well as an architecture similar – or even identical – to that of production.

9.14.3. *Coverage*

The measure of the expected coverage can be defined in the requirements according to the criteria desired by the users (e.g. reliability rate, MTBF, etc.). Often there is no applicable coverage calculation.

9.14.4. *Limitations and risks*

You can use the following checklist to identify the risks and limitations of using this technique:

– Has the expected frequency of use been described at the functional level?

– For each type of user, is it described which functions can be performed?

– Is it described, for each function, when it will be used (daily, weekly, monthly, annually, during the day, in the evening, etc.)?

– Is there information about the relationships between batch procedures within (sub)systems?

– Has the configuration of the production environment been described (hardware, network, OS, DBMS, settings, etc.)?

– Have specific requirements been defined for the performance of online features?

– Is there a distinction between the response time when a function starts and when the screen is refreshed following this function?

– Have performance targets been set for data retrieval and manipulation?

– Have performance targets been set for batch processes?

– Have performance targets been set for memory usage?

– Have requirements been set for the number of database calls that can be made per transaction?

– Are there requirements for maximum page and buffer sizes?

– Are there any requirements for maximum application and/or database size?

9.15. Other types of tests

Some tests can be carried out at various levels or be transversal and cover several levels. These are not testing techniques per se, but we include them here for the record.

9.15.1. *Regression tests or non-regression tests (NRTs)*

Any modification can have an impact (a side effect, a regression) on other parts of the application or the system. These regressions can be of three types:

– local, when a modification in a component generates an anomaly in the same component;

– exposure, when a modification in the component exposes another anomaly that did not appear previously, in the same component;

– remote, when a modification in one component generates an anomaly in another component; this can be the case when data is modified in the component and used in another part of the system.

You should ensure, at each change and for each test level, that no undesirable side effects (i.e. regression) are introduced. Often regressions are identified not by unit testing, but during integration testing or system testing. It is therefore necessary to consider rerunning all existing tests on the application if we wish to identify potential remote regressions. An impact analysis, for example, from the system architecture, can help identify remote regressions to consider.

The frequency of changes (evolutions and corrections) makes the frequent execution of manual regression tests inefficient. It is therefore recommended to:

– define a subset of the tests to be rerun as regression tests and automate them in order to detect local regressions and exposures (e.g. automatic launch at each build). Early detection is thus ensured, which allows rapid correction;

– limit the frequency of changes in order to limit the need to rerun regression tests, since these regression tests are often time-consuming and expensive and their effectiveness is limited (according to Capers Jones, approximately 5% of defects come from regressions);

– perform regression tests only on a stable element of validated quality. This item should be considered a valid candidate for delivery. This solution generates – after the design period – a verification period which is not present in the recommendations of agile methods, but makes it possible to ensure the synchronization of the component, product or system with the other components, products or systems of the system-of-systems.

Terminology: we speak of "non-regression tests" and of "regression tests". It is the same thing as unwanted side effects are called regressions since features can regress from previous operation. In "regression" testing, we test to identify regressions, while in "non-regression" testing, we make sure that there are no regressions.

(Non)regression testing – because it reruns the same tests each time the code changes – lends itself well to the use of test automation tools, to quickly replay a large number of tests in a given – reduced – time period.

– Usefulness: the number of defects found by (non)regression tests is generally very low, and the effort required to run these tests is relatively high. As unit tests and integration tests focus on the operation of each component and its interfaces, the identification of impact on previous functionalities is not considered. The execution of NRTs makes it possible – by identifying the impacts on previously developed functionalities – to limit the risks associated with modifications before the production of corrected versions.

– Impact: given the investment required to ensure the absence of regressions and the very short deadlines before implementing the modifications or developments, the (non)regression tests are often quite limited. This results in a small but steady increase in defects in applications over their lifetimes. Whatever the development cycle chosen, the (non)regression tests must ensure that the developments and changes have not introduced any side effects on the components developed previously. It will therefore be interesting to automate these tests as new features are developed, in order to run them regularly without generating test debt (here a set of features without automated NRTs).

9.15.2. *Automated tests*

Automated testing: any testing task can – more or less easily – benefit from the use of tools. We mainly have the management tasks (monitoring, reporting, management of anomalies, etc.) and the tasks of executing scenarios and test cases. Many tools exist, each focusing on one – or several – aspects of test activities. We mainly consider the activities of recording and executing test scripts, as well as test reporting activities:

– Usefulness of test execution tools: the periodic execution of (non)regression tests justifies the use of tools that can rerun test cases, simulate user actions and ensure that the results obtained correspond to the expected results. The advantage of such tools is to limit the execution time, or even to allow the execution of a large number of tests with a reduced number of testers. The implementation of a test execution tool should be considered as a development project in its own right and not neglect the reporting aspects of these executions. In addition, the use of test automation tools during system and acceptance testing phases on a system-of-systems and hardware equipment may require the design of specific tools, which increase the costs and complexity of these tests.

– Usefulness of test reporting tools: the execution of a test brings value to the project only when it allows decisions to be made about the project. It is therefore essential for the test project manager to provide information (evidence) that the specified contractual requirements have been covered by successfully passed test cases. Thus traceability from requirements to the execution of tests on a known version of the software components is essential.

– Impact of test execution tools: if we limit ourselves to activities of capturing and replaying test scripts, these automation tools have only a limited utility: the investment necessary for the design, the development and maintenance of test scripts is very high and the number of defects identified is limited. On the contrary, using these tools in the context of data-driven testing or keyword-driven testing makes it possible to reuse the same scripts in a very effective and efficient way by giving a certain level of intelligence to the test scripts.

– Impact of test reporting tools: progress statistics, trend curves allow the test project manager to monitor their project and anticipate the actions to be implemented to limit the risks. Many tools do not provide this type of information, but allow us to export data that will be more easily analysed and formatted by spreadsheets.

9.15.3. *Performance tests*

Performance often depends on the product architecture and improving it often requires product re-design. Performance analyses and calculations, based on prototypes, make it possible to anticipate the product's ability to achieve the desired performance (this definition is sometimes called budgeting). To ensure that this performance is achievable, it is recommended to perform performance tests from the lower test levels (the smallest granularity). This makes it possible to check the appropriateness of the budgeting of the individual transactions.

9.15.4. *Security tests*

The objective here is to identify possible security vulnerabilities present in the software. For this you will need to analyse the vulnerabilities present in OWASP or any other list of applicable vulnerabilities. However, this solution does not replace a detailed analysis by the architects and designers, who are often the only ones to know and master these aspects of the software. A technical watch on the publication of "zero day" vulnerabilities and a verification of their applicability or not is also necessary, given the rapid evolution of threats by governmental entities (e.g. China, Russia, USA, Israel, etc.) or not.

Be careful, if a security flaw is detected in one version, it must be checked for its presence – or not – in all the other versions of the software, then ensure that all the versions are corrected simultaneously.

Several repositories exist for IT security testing, including CompTIA, OWASP, NIST, ISTQB Advanced Level security tester, etc. However, as computer security aspects evolve very quickly, we suggest that you check for applicable updates.

According to OWASP, the solution to test IT security of systems after they have been created is inefficient and subject to prohibitive cost. A more adequate solution would be to prevent the introduction of security flaws by including security aspects in each of the design phases. Testing techniques recommended by OWASP are:

– manual inspections and reviews;

– threat modeling;

– code reviews;

– penetration tests (pentest).

To protect your organization from ransomware attacks, the NIST (National Institute of Standards and Technology www.nist.gov) recommends:

– using antivirus software configured to automatically scan emails and removable media (CD, USB storage, etc.);

– keeping all machines up to date with security patches;

– using software products or services that block access to known ransomware sites on the Internet;

– configuring operating systems to use only authorized applications;

– restricting or prohibiting the use of personal equipment on the company network, and in the case of teleworking or remote work, taking additional measures to ensure security;

– whenever possible, using standard user accounts rather than accounts with administrator privileges;

– avoiding the use of personal applications and websites such as emails, chats and social media on work computers;

– not opening files, clicking on links, etc. from unknown sources without first checking them for content (e.g. running an antivirus on a file and carefully inspecting the links).

With systems-of-systems involving interactions between many organizations, it is likely that you will experience many daily threats. It is therefore recommended to set up a recovery plan in the event of a disaster, to develop it meticulously, to implement it and to test it periodically, as well as to maintain an up-to-date list of internal and external contacts to be informed.

In systems-of-systems, security aspects also include the protection of intellectual property and confidential information in terms of performance, level of resistance and resilience, etc. Indeed, the co-contractors on the system-of-systems project can be competitors on other projects and the commercial and industrial data protection aspects are very important. Of course, as with computer security, vulnerabilities are mainly to be seen on the human side.

9.16. Combinatorial explosion

One of the most important aspects in the field of testing is the need to process in a finite time an extremely large number of test cases, almost infinite, resulting from a combinatorial explosion of the various factors and the values that these can take.

For example: 10 factors each having 17 possible values will generate 2,015,993,900,449 combinations, which is clearly impossible to process in a reasonable time. It is therefore necessary to reduce this combination. Two possibilities are available:

– use orthogonal tables that will reduce the number of test cases to less than 289, with equivalent coverage of the values of each factor (without favoring one or the other of the factors); or

– use classification trees that will reduce combinatorics by favoring one or more factors (those that will be tested more strongly).

Reducing the number of tests will increase the risk that defects will remain in the software. It will therefore be necessary to assess the level of risk before reducing the test coverage.

9.16.1. *Orthogonal array testing (OAT)*

Orthogonal arrays (Orthogonal Array Testing – OAT) are mathematical constructions (matrices) where the factors are present in an equivalent way. There are several sources of orthogonal arrays on the Internet (see the works of Neil J.A. Sloane[1] and of the University of York[2]).

For test purposes, it is possible to select combinations of the two highest factors (one factor per type of variable data) and to select the table that will cover all the factors to be processed.

Example: we have below a set of 17 management rules (17 factors) with options ranging from 2 to 7, for a total of 69,672,960 combinations:

– two options: RG03, 04, 06, 07, 11, 12, 13, 16 (256 combinations);

– three options: RG02, 08, 10, 15 (81 combinations);

– four options: RG09, 17 (16 combinations);

– five options: RG01;

– six options: RG05;

– seven options: RG14.

We will therefore need a table with at least 42 rows, 17 columns including one varying from 1 to 7, one from 1 to 6 and one from 1 to 5. There is an orthogonal table offering 18 columns varying from 1 to 17, for 289 lines. This table will be able to meet our needs, so we will have reduced our effort to 289 test cases. All we have to do is replace the identical values in a column with the options of the management rules, and this for the first 17 columns. The 18th column can remain empty and unused.

Attention: this reduction of the combination does not guarantee the coverage of all the combinations, but only that we will have covered – in an equivalent way – the combinations of values to be tested.

---

1 http://neilsloane.com/oadir/.

2 https://www.york.ac.uk/depts/maths/tables/orthogonal.htm.

9.16.2. *Classification tree testing (CTT)*

Techniques to reduce combinatorial explosion often consider combinations to be of equal importance. This is not the case in reality: "happy paths" are more frequently used than exception paths, even if the latter can generate problematic or faulty behaviors. It may be reasonable to test by adapting the frequency of appearance of a variable or a combination of variables.

The classification tree technique defines the variables and their values to allow coverage of each variable value and of the combination of variables, by allowing certain combinations or values to be privileged. Referring to Figure 9.5, some values can be grouped together such as Strikethrough and Double Strikethrough, Uppercase and Small Caps, or Exponent and Subscript. Each of these pairs actually represents a triplet where one of the two boxes can be checked but not both simultaneously. A classification tree can be designed (see Figure 9.6) which defines the test cases to cover the desired combinations. We note that at this stage of designing the classification tree, the (typical) size is the only one that is treated. It will be necessary to process Small and Large as well. It should be noted that in general the setting of texts in MS Word is mainly focused on a normal size, so we can emphasize this font size.

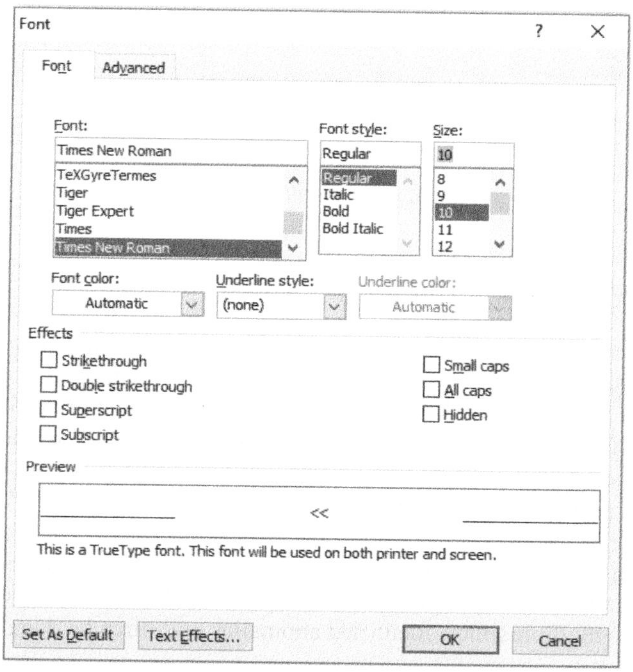

Figure 9.5. *MS Word font options*

As you can note, this technique is greatly helped by tools (here the TESTONA tool) which allows us to view the various branches (grouping of options), their options and the level of coverage achieved. It is clear that if the classification tree becomes too large (too many branches and options), the representation will be less easy to analyse.

The advantage of this technique is therefore that it allows the test cases to be focused on certain combinations. Of course, this is at the expense of other combinations which will not be tested and could therefore contain latent defects.

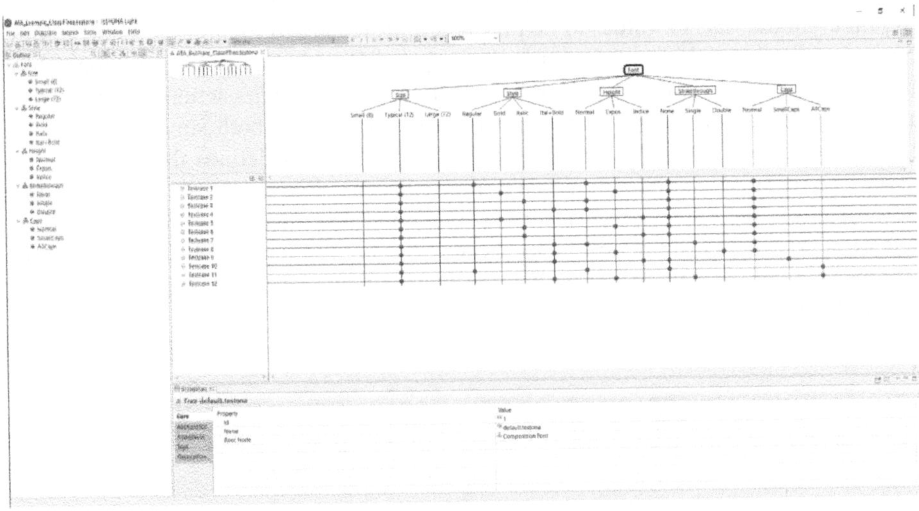

Figure 9.6. *Classification tree (partial)*

9.16.3. *Domain testing (DOM)*

This domain testing technique (DOM) combines the technique of testing EPs and the test of limit values, applied to juxtaposed EPs of numerical type. The term "domain" is to be understood here as a subset of variables that are processed by the system under test and for which it is possible to determine whether their values are in an interval comprised by limit values. We will focus on limits with ON and OFF options and on partitions with IN and OUT values.

The guiding principle of domain testing is to consider the values within each partition, the values below, above, and on each of the boundaries, avoiding testing the same things multiple times. Identified anomalies will often be due to incorrect or ambiguous requirements. According to Beizer (1995), about 30% of anomalies come from requirements, and domain testing would identify 5% of all anomalies.

The domain test is explained in detail by Beizer in the book cited above.

9.16.4. *Built-in tests (BIT, IBIT, CBIT and PBIT)*

There are several types of activities which are not tests in the sense of finding faults in the system under development, but which are looking for failures in the system-of-systems during its execution, the built-in tests (BIT) of which:

– PBIT: Power-on Built-In Tests;

– CBIT: Continuous Built-In Tests;

– IBIT: Initiated Built-In Tests.

These tests are very frequently used in military and aeronautical systems, but you find them in almost all systems-of-systems such as vehicles, for example. On the dashboard of your vehicle, when you switch on the ignition, you have a whole series of tests which are carried out with information which is displayed to inform you of the ability – or not – of the vehicle to start. These are PBIT, tests that are performed when the system is turned on. When driving, you are warned if the temperature of your vehicle rises too quickly or if the oil pressure drops or if the fuel level drops below a certain limit. Here, we will talk about CBIT, continuous testing. In some cases, you have the option of pushing a button to ensure the operation of certain indicators, so we will talk here about IBIT, initiated tests.

These built-in tests are intended to assess the ability of the system-of-systems to function properly. Their purpose is not to detect failures in the system during its design, but during its use. In information systems, complex systems and systems-of-systems, monitoring the operation of systems makes it possible to measure, identify trends and act quickly if necessary. This keeps the system in optimal condition and prevents fatal failures.

The implementation of BIT (Built-In Tests) by measuring and displaying data showing the state of an application or a system can be envisaged within the framework of software quality monitoring, including in the framework of agile development modes such as DevOps: the results of static code analyses, execution of component tests and "smoke tests" can be integrated in order to give a synthetic view of the quality of the software application evaluated.

10

Static Tests, Reviews and Inspections

Defects are the result of human error (or mistakes), static in a deliverable that – upon execution – generate an unintended result. We can have defects that are introduced during the software design phases, and also during the project phases. In the software design phases, defects can occur in (non-exhaustive list):

– the requirements elicitation phase, when a requirement is not identified or misunderstood;

– the phase of writing the requirements, if it is incorrectly filled in (typo error);

– the software analysis and design phase, if the model or architecture is not correctly defined or if the interfaces between the components are incomplete;

– the coding phase, if the understanding of the requirements and their translation into code is incorrect;

– the component design phase;

– the test design phases – whatever their level – based on the requirements and selected techniques.

During the phases of a test project, defects can appear in:

– the needs analysis phase to design a test strategy, if the quality requirements are not described correctly;

– the preparation phases of the test environments, whether in terms of the necessary data, the applications with which the software product must communicate or the hardware requirements;

– the test data definition phases, whether input data, configuration and/or parameterization data, or expected data that will be compared with the data obtained during the tests;

– the results analysis phases, where it is possible to identify a lack of test coverage in relation to the coverage objectives, or a lack of effectiveness of the test techniques used in relation to the announced objectives of the techniques used.

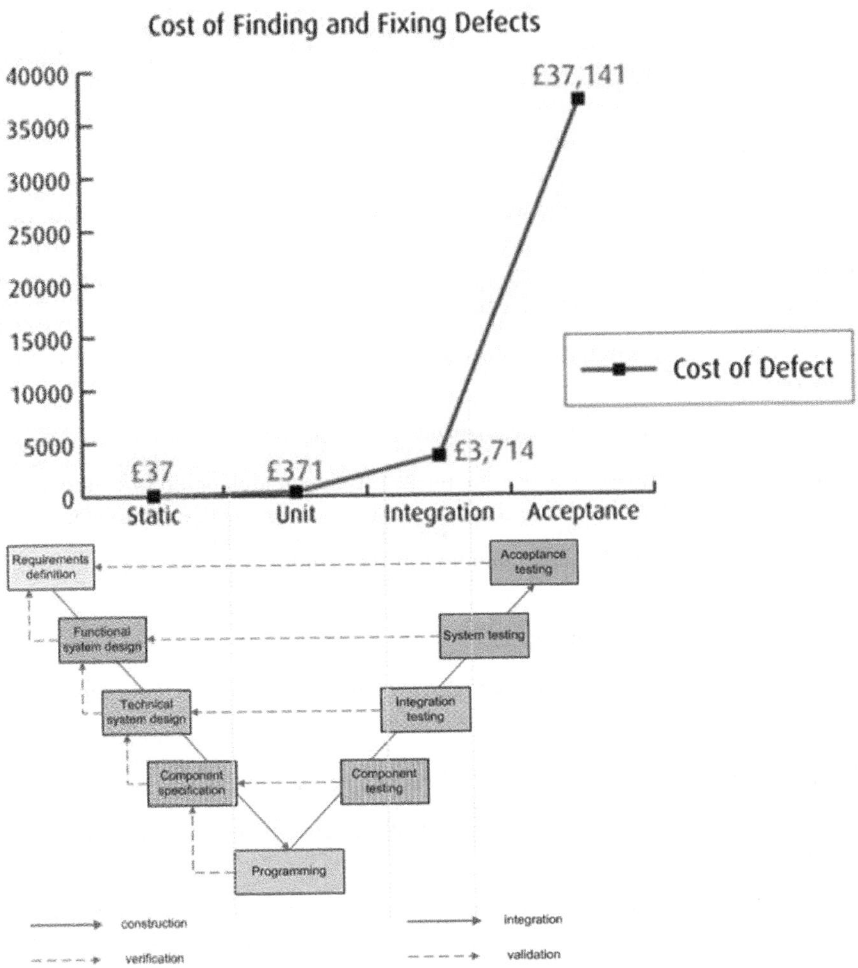

Figure 10.1. *Shift left and costs identification*

At each stage of code design, deliverables from previous phases are used as input to the current phase and new deliverables are created at the end of this phase. The defects introduced in one phase serve as a basis for – erroneous – developments in subsequent phases. The cost of defects therefore increases if they are detected in phases subsequent to their introduction phase.

It is common to find defects in documents, whether they are spelling mistakes, typos, logical mistakes or other types of defects. A quick reading can often detect these faults. Further reading can find more subtle flaws. This type of detection does not require designing test plans, test cases or test scenarios – automated or not.

The use of reviews allows early and rapid detection of defects: there is no need to design tests to highlight the defect, and the development team has not yet designed the code. The formalism of the journals ranges from totally informal to very formal. The effectiveness of the reviews varies according to the knowledge and skills of the participants, and also to the formalism of these reviews.

Reviews were proposed by Michael E. Fagan to IBM in 1976 and have proven their efficiency and cost-effectiveness ever since. Unfortunately, they have not yet fully entered into the current practices of development teams.

10.1. What is static testing?

Static testing and reviews are component verification activities that do not require those components to be running. Static techniques can use tools (we will talk about static analysis) or individuals (we will talk about reviews). Formalism and level of detail vary resulting in a variation in the effectiveness of these activities.

Review activities can and should be present at each level of testing, both on test input and output, and also on processes and report generation.

– Usefulness: the effectiveness of reviews has been demonstrated since the 1970s and varies from 50 to 85% of fault detection. These activities can be carried out even before the software is executable, therefore very early in the delivery, which leaves more time for corrections.

– Impact: the early detection of faults, including design faults, saves a lot of time and increases efficiency.

– Objects: the elements that can be subject to reviews are many and varied as defined in IEEE1028 (Standard for Software Reviews and Audits 2008) and ISO20246 (2017). For static tests, we will mainly have all the documentation (e.g. Specifications, Expressions of Needs, Requirements, User Stories, etc.), architecture elements, and also workflows, processes and plans of project.

– Repository: the static analyses will compare the components with respect to a repository specific to the analysed language. The reviews will compare the objects against reference documents (e.g. Specifications, Expressions of Needs, Requirements, User Stories, etc.).

– Techniques: the various types of reviews and their formalism are described in IEEE1028 (Standard for Software Reviews and Audits, 2008) and in ISO/IEC20246 (2017). The effectiveness of reviews varies according to the formalism, from the most formal to the most effective.

10.2. Reviews or tests?

A debate exists on the distinction between reviews and testing. Some consider it not testing, while others consider reviews to be an integral part of testing.

If the definition of testing is (systematic) defect finding, then any systematic defect detection technique should be considered testing, including reviews. In fact, reviews are early testing activities that prevent defects from spreading to subsequent phases.

10.2.1. *What is a review?*

A review is a comparison of a written document with reference data (formalized or in the minds of experts), with the aim of identifying defects or deviations from this reference system. This comparison is made by individuals with a variable level of formalism. Since reviews can be performed before code is designed, they are considered the most efficient and cost-effective testing techniques.

The main difference between a review and a static analysis is that static analyses are performed with tools whereas reviews are performed by individuals.

10.2.2. *What can be subjected to reviews?*

In general, it is possible to submit to a review process any document that is readable or understandable by a human being, whether it is a description of a process, a document or the results of a test campaign, graphics and diagrams of organization, etc.

In the context of this work, we will focus the targets of the reviews mainly on:

– any input data, necessary for the execution of the planned tasks. In the agile world, it is often associated with the concept of "DOR" for Definition of Ready, which is a checklist defining all the data and information necessary for a task to be executed without having to wait for information not yet available from other activities;

– output data, defined in the agile world under the concept of "DOD" for Definition of Done, which is a checklist defining from when a task can be considered completed. Often the DOD mentions updating documentation, providing test results, etc.;

– processes, which can also be reviewed to ensure their effectiveness and efficiency. Analyzing task-defining processes helps identify input data needs, ensure output data is properly planned, measured and produced, check if resources are available and if people need the trainings;

– other documents, which may also be subject to review, such as contracts, design specifications, development plans, maintenance plans, program source code, project plans, requests for proposals (RFPs), requirements specifications, plans manuals, test specifications, user manuals, user procedures, etc.;

– code, architecture, documentation, user stories and manuals, which can also be reviewed.

We therefore understand that it is possible to submit almost any deliverable to a review process.

10.3. Types and formalism of reviews

Standard ISO20246 (2017) offers various types of reviews, each with its own attributes, objectives and roles. We summarize them below. The value "M" implies an obligation (mandatory), the value "O" implies an option, and the absence of value indicates that the information or the objective is not material.

The level of formalism and rigor will vary depending on the product to be reviewed: if the risk associated with the work product to be reviewed is high, the review should be formal and rigorous. For milestone reviews, it is recommended to include all documents and deliverables to inform non-technical stakeholders of the process and obtain their agreement.

Objective	Types of review								
	Author verification	Verification by a colleague	Informal group review	Inspection	Milestone review	Peer review	Colleague review	Technical review	Technical walkthrough
Discover problems	M	M	M	M		M	M	O	M
Evaluate quality and increase confidence in the product				M	M			O	O
Training							O		O
Obtain a consensus					M			M	O
Generate new ideas		O	O			O	O	O	O
Motivate and enable improvement				O			O	O	O
Reduce time, cost and risks	O								

10.3.1. *Informal or ad hoc reviews*

Implementing a formal process increases the cost of performing reviews. Any process identifying defects – without execution of the application subject to review – in a pragmatic and non-formal way can be considered before setting up a formal review. This will prevent formal reviews from finding too many defects and thus limit costs.

As informal reviews are by definition non-formal, there is no predefined process. There are, however, a number of items that may apply as mentioned in Freedman and Weinberg (1990).

Earlier the better: it is more useful to perform several informal reviews at various stages of software product development to avoid starting a review when an "unresolvable" defect is identified.

Well-defined roles: as with more formal journals, the roles of author, reader/reviewer, leader are important and should be properly defined.

Should informal reviews be done before or after formal reviews? If the decision has been made to implement formal reviews, the authors will prefer to have a first – informal – reading by their peers. So do not be offended by informal reviews before formal reviews.

What are the main differences between informal reviews and formal reviews? Informal reviews are often done with close colleagues of the author "as a family", so the author will have less stress. The results are not published outside the circle of participants and there will be no presumption of guilt. In a formal review, it is preferable that the component under review be of good quality in order to avoid wasting the participants' time.

10.3.2. *Technical reviews*

Technical reviews focus on the technical content of a project, not the context of that material in the business environment. Technical reviews answer the question "will the product do the job requested?" Here the term "product" can be replaced by "component", "equipment", "subsystem", "system", "software", "architecture", etc. The result of a technical review will be twofold; on the one hand, it must inform about the quality of the product subject to review, and on the other hand, it will provide information to management and to the project.

Technical reviews can be formal or not have any formalism and a product can undergo several technical reviews during its development. It is important to note that technical reviews can cover both functional aspects (software features, hardware characteristics of the equipment, etc.) and also non-functional aspects (response time, user interface, types and number connectors, etc.).

In the case of large products, it is tempting to proceed by analyzing samples. This solution is only valid if the samples are representative. According to Freedman and Weinberg (1990), an error costing approximately $235,000,000 was traced to a forgotten system piece of the sampling procedure. The specifications had been reviewed but only on the sample.

10.3.3. *Checklist-based reviews*

It is a systematic approach to identify issues based on checklists. If the various participants in the review are assigned different checklists, this solution offers broader coverage and avoids the duplication of effort inherent in ad hoc reviews.

One of the disadvantages of this technique is that participants will be limited to the content of the checklists. A second disadvantage relates to the size of the checklists: the longer and more exhaustive, the more difficult their implementation. One way to implement a checklist-based review is to find, for each item on the checklist, one or two defects, then say that "the product has defects of this type, with the following examples as examples ...", then let the author or the team of authors correct the whole document on the basis of these typologies of defects.

Of course, nothing prevents improving or completing the checklists by adding additional typologies based on defects that have been found elsewhere. This continuous improvement of checklists is important, both for sequential development cycles and for agile development cycles (see ISO20246 2017).

10.3.4. *Scenario-based reviews*

Certain deliverables (e.g. requirements, design or tests) may be appropriately documented so that a scenario-based approach is most appropriate. Participants check the deliverable to identify if the functionality is correctly described and typical failure conditions are well addressed.

It should be noted that this technique is limited to documented scenarios, so omission flaws – where a required functionality is missing from the deliverable – will not be identified.

Technical reviews are often based on this type of scenario-based review.

10.3.5. *Perspective-based reviews (PBRs)*

According to ISO20246 (2017), this type of review is the most efficient and cost-effective form of defect detection in formal reviews. In this type of reviews, the participants take the point of view of each stakeholder and revise the deliverable from that stakeholder's point of view. This way of doing things avoids duplication of effort between participants.

Typical viewpoints are those of the business analyst, process owner, designer, maintainers, operations, programmers, testers, managers and regulators, and users (depending on their levels and responsibilities).

10.3.6. *Role-based reviews*

This type of review is similar to perspective-based reviews, except that the same participant can successively take on several roles and therefore several perspectives.

10.3.7. *Walkthrough*

Walkthroughs differ from the technical review in that the author presents the element submitted for review. The author is therefore directly available to answer questions and note comments.

10.3.8. *Inspections*

Inspections are the most formal reviews, led by a moderator trained in this type of review. Inspections are characterized by:

– a kick-off meeting where the elements subject to inspection, the participants as well as the duration of the individual work periods and the date of the inspection review are defined;

– a period of individual work, where each of the reviewers will analyse the item under inspection, identifying – and optionally transmitting – the problematic points and subject to questions;

– a formal inspection meeting, where questions, problems or remarks are studied and categorized in order to be provided to the author for correction;

– a decision on the fate of the item subject to inspection, where the participants decide whether to accept (with or without reservation) the item subject to inspection, or whether corrections are requested and whether – after these corrections – the element must (or not) be re-inspected.

The moderator should report to the project management the metrics and measures that will allow evaluation of inspection effectiveness or a need for improvement.

10.3.9. *Milestone review*

A milestone review is a management review and makes it possible to monitor the progress of the project, specifically to check whether all the expected deliverables have been provided with the required level of quality. In systems-of-systems projects, the reviews relate to all the milestones of the project, from the highest level (system-of-systems level) to the lowest level (equipment, software, product). In general, reviews are formalized and often associated with payment milestones.

10.3.10. *Peer review*

Any review can be a peer review in that it suffices that the people who participate in the review are "peers" of the author of the product under review. Care must be taken not to confuse this meaning of "peer" with the meaning of "pair" as in the concept of programming by "pairs" where a couple of programmers share the same keyboard and the same screen to produce – together – a software product.

10.4. Implementing reviews

Implementing efficient and cost-effective reviews requires identifying when in the life cycle the defect is introduced, and the root (initial) cause of the defect. Positioning fault detection techniques closest to the moment of fault introduction is most effective. As defects can be introduced in each of the deliverables, it will be necessary to position reviews at many places in the process. Other aspects to consider are:

– the number of people to include in the reviews, knowing that the more participants there are, the more chances there are of finding defects, and also the more costs and delays will increase. You will seek to have different profiles to have different points of view;

– the duration of personal work assigned to the study of the product submitted for review. If the time allocated for the review is insufficient, or if the participants are overworked, the time they can allocate to the review is likely to be insufficient.

Hence, you should measure the efficiency and costs of each review, by document type and review type, in order to identify the optimal combinations.

10.5. Reviews checklists

To facilitate the execution of reviews, the use of checklists makes it possible not to forget important points which could be evaded in the urgency of the tasks to be carried out.

Examples of checklists for reviews are available in Wiegers (2010), and some are included, by way of example and illustration, in the following sections.

It is relatively easy to define your own checklists from an analysis of past defects, and it is mandatory to periodically improve the review checklists.

If the document is large and the time allocated to the review reduced, a checklist makes it possible to systematically search in a representative sample of the document and to extrapolate the results to the rest of the document. The defects identified serve as an example in the review report. This solution allows – among other things – to determine the need for a further review (e.g. technical review or inspection).

10.5.1. *Reviews and viewpoint*

Comments that can be made about a document submitted for review – for example, a document describing requirements or user stories – will vary depending on the point of view from which this information is analysed. In addition, comments will vary depending on the impacts.

Views may vary:

– as a customer or principal, we could be tempted to mention performance (e.g. response time, reliability, etc.) beyond what is strictly necessary for the needs of the system or the system-of-systems. This way would make it possible to anticipate future needs, without having to define them for the current system;

– as a supplier or MOE, having ambiguous or untestable requirements is interesting because it allows us to deliver only what meets the non-ambiguous aspects and therefore limits design and development costs;

– as a quality specialist or tester, the objective will be to determine how to verify or validate the requirements or User Stories. We could use the acronym IADT where the I stands for Inspection (validation by inspection or review of the product),

A stands for Analysis (analysis of the architecture and design of the product), D stands for Demonstration (by unit tests, integration or system, as well as the use of prototypes) and T stands for Tests (use of the system or system-of-systems during a period of acceptance testing, pilot phase, etc.).

Other points of view may be considered, such as that of system users, system operators or administrators, etc.

The typology of comments will vary from formal (e.g. incorrect document format or typeface) to very complex (requiring a thorough understanding of the expected behavior of the system, its ecosystem or within the system-of-systems), passing by the grammatical aspect (absence of verbs, singular vs. plural, etc.) or non-compliance with contractual or normative obligations.

10.5.2. *Checklist for specifications or requirements review*

This checklist is copyright © 2001 by Karl E. Wiegers.

Organization and completeness

☐ Are all internal references to other requirements correct?

☐ Are all requirements written down to the correct and appropriate level of detail?

☐ Do the requirements form an adequate basis for the design?

☐ Is the implementation priority for each requirement included?

☐ Are all external hardware, software and communication interfaces defined?

☐ Have the algorithms intrinsic to the functional requirements been defined?

☐ Do the specifications include all known customer or system needs?

☐ Is the expected behavior documented for all anticipated error conditions?

Exactness

☐ Are any requirements duplicated or in conflict with another?

☐ Is each requirement written in clear, concise and unambiguous language?

☐ Is each requirement verifiable by testing, demonstration, review or analysis?

☐ Is each requirement within the scope of the project?

☐ Is each requirement free of content error or grammatical error?

☐ Is necessary information missing from a requirement? If so, is it identified by a TBD?

☐ Can all requirements be implemented with known constraints?

☐ Are the specified error messages unique and understandable?

Quality attributes

☐ Are all performance targets specified correctly?

☐ Are all security and safety considerations properly specified?

☐ Are other objectives in terms of relevant quality attributes explicitly documented and quantified, with acceptable trade-offs specified?

Traceability

☐ Is each requirement uniquely and correctly identified?

☐ Is each software functional requirement traceable to/from a higher level requirement (e.g. system requirement, use case)?

Special issues

☐ Are all the requirements real requirements and not design or implementation solutions?

☐ Are all functions with response time criticalities identified, and duration criteria specified for them?

☐ Have the internationalization aspects been adequately addressed?

10.5.3. *Checklist for architecture review*

This checklist is copyright © 2001 by Karl E. Wiegers.

Structure

☐ Does the architecture allow for the implementation of all requirements?

☐ Has the architecture been adequately broken down?

☐ Have the system functions been correctly assigned to the components?

☐ Does the architecture provide a sufficient basis for the design work that will follow?

☐ Can the architecture be implemented?

☐ Have the maintainability aspects been adequately addressed?

☐ Can the programs be integrated and tested incrementally?

Exactness

☐ Does the architecture avoid unnecessary redundancy?

☐ Have all reliability and performance requirements been addressed?

☐ Have all safety considerations been addressed?

☐ Does the architecture consider all existing constraints?

☐ Are all data structures defined, and only those that are needed?

☐ Will the proposed architecture meet all specified quality criteria and performance objectives?

Standards and traceability

☐ Have all architectural standards been followed?

☐ Can all architectural elements be traced to requirements?

☐ Are standard, non-proprietary components used wherever possible?

Logic

☐ Is there missing or incomplete logic?

☐ Have all states and cases been considered?

Interfaces

☐ Are all interfaces clear and well defined?

☐ Are we only passing the minimum amount of data at each interface?

☐ Is the minimum global system data added or impacted?

Clarity

☐ Is the architecture, including data flows, control flows and interfaces, clearly represented?

☐ Are the multiple representations of the design consistent with each other?

☐ Have all decisions, dependencies and assumptions for this design been documented?

10.5.4. *Checklist for high-level design review*

Structure

☐ Has the design been broken down to a point where the next phase can begin?

☐ Has each component been fully specified and testable?

☐ Is the modular breakdown consistent with local standards for modularity and strong coupling?

Data

☐ Have global constants and past data been parameterized?

☐ Have all data been set correctly?

☐ Do the names of data structures and elements respect naming rules and are they understandable?

Correctness and completeness

☐ Does the high-level design implement all requirements?

☐ Has all the infrastructure (backup, restore, synchronization points, etc.) been addressed?

☐ Have all quality attribute and performance requirements been designed?

☐ Have internationalization and translation issues been properly and adequately addressed?

☐ Have treatment priorities been addressed?

☐ Have economic and feasibility analyses been performed to determine that the required performance and technical objectives will be achievable with this design?

☐ Is there a memory size estimate to estimate storage requirements for each module, table and file?

☐ Have all security considerations been designed?

☐ Does the high-level design consider all existing constraints?

☐ Is the high-level design unnecessarily redundant?

☐ Have the maintainability aspects been adequately addressed?

☐ Is the high level design complete, correct and unambiguous?

Standards and traceability

☐ Have all design standards been followed?

☐ Does the human-machine interface follow project standards?

☐ Can all parts of the high-level design be traced back to requirements?

Interfaces

☐ Are all interfaces clear and well defined?

☐ Is the minimum amount of data passed through each interface?

☐ Is the minimum global system data added or impacted by design?

Robustness

☐ Have self-test requirements, safety devices and degraded modes been planned and detailed?

☐ Has an error handling mechanism been identified?

☐ Are all unusual situations handled in a reasonable and non-destructive manner?

10.5.5. *Checklist for critical design review (CDR)*

This checklist is copyright © 2001 by Karl E. Wiegers.

Structure

☐ Is the pseudo-code or representation format consistent in terms of level of detail?

Data

☐ Have all data been correctly defined and initialized?

☐ Are all defined data used?

☐ Are the names and types of the elements in accordance with the project's data dictionary?

☐ Are default values used, and are they correct?

Exactness and completeness

☐ Is the detailed design a correct and complete implementation of the high-level design?

☐ Are the external specifications of each module complete and testable?

☐ Have all digital techniques been analysed for accuracy and precision?

☐ Have critical timing aspects been analysed?

☐ Has the high-level budgeting for memory been expanded and detailed and updated?

☐ Have all functions been clearly specified and are they logically independent?

☐ Have maintainability concerns been properly addressed?

☐ Does each module have a high internal cohesion?

☐ Does each module have weak external coupling?

☐ Is the detailed design verifiable?

☐ Is the logic correct, clear and complete?

☐ Have all user interfaces been completely designed?

☐ Can the end conditions of the loops be achieved?

☐ Can the logic be tested?

Standards and traceability

☐ Have all detailed design standards been followed?

☐ Do call protocols follow project standards?

☐ Are all parts of the detailed design traceable to the high level design and requirements?

Robustness

☐ Are error conditions handled non-destructively?

☐ Can corrective actions be taken by the module that identifies the error?

☐ Are unusual conditions handled in a reasonable and non-destructive manner?

10.5.6. *Checklist for code review*

Structure

☐ Does the code fully and correctly implement the design?

☐ Does the code comply with relevant coding standards?

☐ Is the code well structured, and is the style consistent in style and format?

☐ Is there unreachable dead (uncalled) code or useless procedures?

☐ Are there any remaining stubs or test routines in the code?

☐ Is there code that can be replaced by reusable external calls or library function calls?

☐ Are there repeated blocks of code that could be condensed into a single procedure?

☐ Is memory usage efficient?

☐ Are symbolic variables used rather than numeric or textual values, constants?

☐ Are there overly complex modules that should be restructured or split into multiple routines?

Documentation

☐ Is the code clearly and adequately documented with a style of comments that is easy to maintain in the code?

☐ Are all comments consistent with the code?

Variables

☐ Are all variables properly defined with clear, consistent and understandable names?

☐ Are all assigned variables type consistent or correctly cast?

☐ Are there redundant or unused variables?

Arithmetic Operations

☐ Does the code avoid comparing floating point numbers for aspects of equality?

☐ Does the code systematically prevent rounding errors?

☐ Does the code avoid adding or subtracting numbers with large orders of magnitude differences?

☐ Are the dividers tested for zero values or noise?

Loops and branches

☐ Are all loops, branches, and logical constructs complete, correct and properly nested?

☐ Are the most common cases tested first in chains of IF-ELSEIF?

☐ Are all cases covered in IF-ELSEIF or CASE blocks, including ELSE or DEFAULT clauses?

☐ Does every CASE order have a DEFAULT?

☐ Are the end-of-loop conditions obvious and still achievable?

☐ Are the indexes or indices correctly initialized, just before the loop?

☐ Can any instruction included in loops be positioned outside these loops?

☐ Does the code in a loop avoid manipulating index variables or using them to break out of the loop?

Defensive programming

☐ Are indexes, pointers and subscripts tested against limits of vectors, tables, records or files?

☐ Are imported data and input arguments tested for validity and completeness?

☐ Are all output variables assigned?

☐ Is the correct data being used in each statement?

☐ Are all memory allocations deallocated?

☐ Are timeouts and error interrupts used for access to external devices?

☐ Is the existence of files validated before trying to access these files?

☐ Are all files and devices left in a correct state in the event of program termination?

10.6. Defects taxonomies

Reviews and inspections make it possible to discover defects in deliverables that are readable and understandable by a human being. This makes it possible to discover defects early, before the code is designed, to correct them before this erroneous information is used as the basis for the development of tests or other activities.

The types of defects can be grouped according to:

– defects related to user interfaces (including defects related to UX User Experience);

– user or programming error management;

– defects related to limit values;

– defects in the calculations;

– defects related to the states of departure and arrival;

– defects associated with control flows;

– defects in the management or interpretation of data;

– race conditions where processes are waiting on each other for response;

– load conditions;

– hardware-related faults;

– defects related to the management of sources, versions and IDs;

– defects related to testing.

In addition to the defect taxonomies that you can extract from the checklists mentioned in the previous sections, Appendix A of Kaner et al. (1999) provides 72 pages of common defect types, which can guide you in designing a defect taxonomy.

A taxonomy of defects makes it possible to carry out groupings and to identify axes of improvement of the quality related to these defects (via the analysis of the root causes at the origin of these defects). Training activities or reminders on the processes and procedures to be implemented can be based on these taxonomies.

10.7. Effectiveness of reviews

Reviews and inspections, as well as static analyses, as they are performed before the code is compiled or executed, allow defects in the code – or in the processes – to be identified at a very early stage. This limits impacts and costs because corrections can be made quickly. This allows substantial savings as defined in Jones and Bonsignour (2012).

To be convinced of the usefulness of reviews, it suffices to think about the time needed to find defects in a document by reading it and compare this to the time needed to analyse the document, design the test conditions, implement the test cases (with the implementation of test data and automation if necessary), wait for the delivery of the component to be tested, execute the test and identify a failure then analyse the root causes of the failure to identify the defect.

The effectiveness of reviews, in terms of identifying defects, varies according to the level of formalism implemented – from the most formal to the most effective – and the level of competence of the participants in these reviews. According to Capers Jones, this quality improvement goes up to 24.70% for requirements inspections. The improvements provided by reviews also relate to team productivity (avoid having to redo tests, correct code, etc.) with an average productivity improvement of 16.88%.

10.8. Safety analysis

The purpose of safety analyses is to ensure that the system (or system-of-systems) does not generate risks to human life. The techniques often used are FMEA (Failure Mode and Effect Analysis), FMECA (Failure Mode and Effect Critical Analysis) and FTA (Fault Tree Analysis). These techniques use fixed values – feedback from the industry – to determine the reliability rate of hardware systems by combining the reliability rates of components with each other. However, the reliability rate of software is often considered equal to 1 (i.e. 100% reliability) which does not correspond to the reality that we know (for more information, see Broekman and Notenbloom (2003), Stamatis (2003), Musa (2004) and Engel (2010)).

Terminology

ATDD: Acceptance Test-Driven Development, a software design methodology based on the design of acceptance tests (unit testing level) and the development of the minimum code necessary for the tests to pass successfully. It is related to BDD and TDD. It should be noted that the term "acceptance" is taken here in the sense of acceptance by the developers and not "acceptance" by the users.

BDD: Behavior-Driven Development, an agile development method combining the principles and techniques of TDD and languages (e.g. gherkin) to define the behavior and the expected results. BDD is considered an effective technique or practice when the problem to be solved is complex.

BI: Business Intelligence, the term encompassing the strategies and technologies used by companies for data analysis and business information management. In general, this involves the analysis of large volumes of data, structured or not, in order to interpret this data, identify opportunities and implement effective strategies based on their analyses.

BIT: Built-In Test, an internal component test (hardware or software) allowing the component to ensure its proper technical or functional operation.

BVA: Boundary Value Analysis, a testing technique focusing on the analysis of boundary values (valid and invalid) of ordered equivalence partitions.

CAF	*Capacité à Faire*, team bandwidth, the ability of a team to perform a job or effort.
CBIT	Continuous Built-In Test, an internal component test (hardware or software) that allows the component to continuously ensure its proper technical or functional operation. Such a test allows the component to warn a supervisor in the event of a drop in performance or other non-operations.
CCB	Change Control Board, a change management committee. This may concern requirements, but, more generally, configuration changes of the systems under its control.
CCP	Conformance Creation Plan, a plan defining how the requirements will be demonstrated as compliant, via a variation including Inspections, Analysis, Demonstrations (software tests) and Tests in operational environments (acceptance tests and pilot phases). This CCP is a stage in the design milestones, close to the start of the design and to be included in the PDR milestone.
CD	Continuous Delivery, a software engineering approach where teams deliver software in short cycles, ensuring that software can be delivered reliably at any time, without doing so manually. This approach tends to reduce cost, time and risk, by enabling incremental application updates. A simple and repeatable process is important for continuous delivery.
CD	Continuous Deployment, a software engineering approach where features are delivered frequently through automated deployments. In an environment where data-driven microservices provide the functionality, and where microservices may have multiple instances, CD consists of instantiating a new version of a microservice and disposing of an old version.
CDR	Critical Design Review, a design milestone review, after the PDR and intended to allow the start of the design of the subsystem or system with the least possible risk.
Certification	A formal attestation or confirmation of certain characteristics through an organization certification process.

CI/CD (Combined) Continuous Integration and Continuous Delivery practices, forcing automation in building, testing and deploying applications. DevOps practices involve continuous development, continuous testing, continuous integration, continuous deployment and continuous monitoring of software applications throughout their life cycle. The CI/CD Pipeline underpins DevOps operations.

COTS Commercial Off The Shelf, an existing commercial product on the shelf, which can be purchased "as is". This applies to both software and hardware products.

Coverage The rate by which a set of requirements or risks are covered by test cases as well as executions of these test cases, in order to demonstrate that these requirements or risks are correctly verified and validated without forgetting any aspect.

CRI Coverage of Risks, a metric for measuring risk levels according to their criticality.

CTT Classification Tree Testing, a testing method based on classification trees.

DCOT Data COmbination Testing, a test method based on the combination of nominal data and their use. It is often necessary to have a diagram of the data architecture in order to identify when and where this data is used in the application or system.

DCYT Data CYcle Testing, a testing technique focusing on the life cycles of data (creation, referencing, use, destruction) within subsystems and systems, requiring an understanding of the architecture of data in systems and subsystems.

DDE Defect Detection Efficiency, the metric to determine the efficiency of processes in detecting defects.

Debugging The process for finding, analyzing and eliminating the causes of failure in software, mainly carried out by designers and developers.

DOD	Definition of Done, a checklist defining when a task can be considered completely finished (which should include updating documentation, providing test results, etc.).
DOM	DOMain testing, a test technique combining the equivalence partitions and the boundary value testing techniques, applied to adjacent ordered equivalence partitions (usually numerical values). The term "domain" is to be understood here as a subset of variables which are processed by the system under test and for which it is possible to determine whether their values are in an interval comprised by limit values.
DOR	Definition of Ready, a checklist defining all the things necessary for a task to be able to run without having to wait for information not yet available from other activities.
DTT	Decision Table Testing, a test technique making it possible to limit the combination of tests to be executed when there are combinations of business rules to be processed.
E2E	End to End, tests that carry out a functional process from the beginning to the end of this process. This type of test can be run on a single system as well as interact with several interconnected systems.
EAI	Enterprise Application Integration, a middleware that integrates systems and applications across a company. EAIs connect applications in a way that simplifies and automates business processes as much as possible, while avoiding significant changes to enterprise applications. This allows data to be integrated between various systems, eliminates dependencies on a vendor and provides a common interface for users.
ETL	Extract, Transform, Load, a generic procedure for copying data from one or more sources to one or more destination systems. These software packages include the conversion of data from one format to another and their transfer.
Fault injection, fault seeding or defect injection	A test technique aimed at verifying that the defined test processes (around automation tools) are capable of identifying the types of anomalies expected. This technique consists of creating known anomalies and checking whether they are detected by the processes and tools implemented.

GQM	Goal Question Metric, a system measurement approach that defines: – a conceptual or managerial level (the Goal); – an operational level (set of Questions) to focus on a specific characteristic; – a quantitative level (Metric) based on the models and associated with each question.
IBIT	Initial Built-In Test, an internal test of the component (hardware or software) allowing during its initial launch to ensure its proper technical or functional operation.
Inspection	A formal review of a work product (deliverable) to identify issues, which uses defined roles and metrics to improve the review process.
MTBF	Mean Time Between Failures, the time between two failures calculated as the arithmetic mean between two system failures. This measurement depends on the definition of a system failure, usually when the system is "out of order". The higher the MTBF, the longer the system will operate before failing.
MTTR	Mean Time to Recover, an average time for the system to recover and restart a normal operation.
MVP	Minimum Viable Product, the version of a product with enough features for customers to use.
NRT	Non-Regression Tests, functional tests aimed at ensuring the absence of regression or side effects between a previous version and the current version.
ODC	Orthogonal Defect Classification, an orthogonal classification of defects. This methodology was developed by IBM in the 1980s, associating defect typologies with design phases/processes and measuring the presence of these defect typologies in order to improve design processes.

PBI	Product (or Project) Backlog Item, element, requirement, user story, use case to be developed or anomalies to be corrected in the Agile Scrum world, which describes what is required of a product (or project), generally sorted in descending order of priority.
PBIT	Power-on Built-In Test, a test carried out during the start-up of a subsystem or a system and making it possible to display the state of the subsystem or system.
PBR	Perspective-Based Reviews.
PCT	Process Cycle Test, test of the process cycle, test method intended to ensure that administrative actions implemented obtain the expected results.
PDR	Preliminary Design Review, a design milestone where the main orientations are defined by the design teams and validated by the client teams.
RCA	Root Cause Analysis, an analysis of the processes in order to identify the initial – root – cause for a defect.
Requirement	A condition or ability required of a component, product or system to solve a problem or achieve an objective that must be held or possessed by this component, product or system in order to satisfy a contract, standard, specification or other formally or informally imposed document. Requirements can be formalized in various documents, for example: User Stories, Use Cases, Features, etc. or be informal (not formalized).
Review	An evaluation of the state of a product or project to identify deviations from the expected results and recommend improvements. For example: management review, informal review, technical review, technical inspection and proofreading.
Root cause	The cause of an anomaly; see also RCA.
RUN	The activity or phase of testing, after SETUP, that actually executes the test.

SETUP — The activity or test phase to prepare the data and environment necessary for the execution of a test that will take place in the RUN phase.

SLA — Service Level Agreement identifies the level of service (e.g. response time, availability, etc.) expected from the system or component. SLAs are often associated with penalties and contractual clauses between suppliers and users.

SMART — Specific, Measurable, Achievable, Realizable and Traceable (sometimes also considered achievable over time), and applying to goals or requirements.

TDD — Test-Driven Development, a software development methodology based on the automation of low-level unit tests. It starts with designing a test that – since there is no code it cannot run successfully – followed by designing the minimum code for the test to run successfully.

TDR — Technical Debt Ratio, a metric to measure technical debt.

TEARDOWN — A test activity or phase that removes test data created during testing to allow test reruns in the future.

Technical review — A formal peer review of a work product by a team of technically qualified people who examine the deliverable and identify deviations from specifications and standards.

Test — The process present in all life cycle activities, static and/or dynamic, concerning the planning and evaluation of software products and related products to determine whether they meet the requirements, to demonstrate that they are fit for the purpose and detect anomalies. This is sometimes known as V&V (Verification and Validation).

Test case — A set of input values, execution preconditions, expected results and execution post-conditions, developed for a particular test objective or condition, such as executing a particular path of a program or verifying the compliance with a specific requirement.

Test condition	Something – a behavior or a combination of conditions – that may be interesting or useful to test.
Test objective	A reason or purpose for designing and performing a test.
Validate	Demonstrate, by providing objective evidence, that a system can be used by users for their specific tasks [ISO29119-1].
Validation	A process to ensure that specifications are correct and complete. The system life-cycle process can use software specifications and derived requirements for system validation [DO178B].
Verification	An evaluation of the products of a process in order to guarantee their accuracy and consistency with respect to the input data and the rules applied to this process [DO178B].
Verify	Demonstrate, by providing objective evidence, that specific requirements have been met [ISO29119-1].

References

Andrews, M. and Whittaker, J. (2006). *How to Break Web Software*. Addison-Wesley.

Bath, G. and van Veenendaal, E. (2014). *Improving the Test Process*. Rockynook.

Beck, K. (2003). *Test-Driven Development by Example*. Addison-Wesley.

Beizer, B. (1990). *Software Testing Techniques*, 2nd edition. Van Nostrand Reinhold.

Beizer, B. (1995). *Black-Box Testing*. Wiley.

Berenbach, B., Paulish, D.J., Kazmeier, J., Rudorfer, A. (2009). *Software & Systems Requirements Engineering in Practice*. McGraw Hill.

Biafore, B. (2010). *Microsoft Project 2010*. O'Reilly.

Black, R. (2003). *Critical Testing Processes: Plan, Prepare, Perform, Perfect*. Addison-Wesley.

Black, R. (2008). *Advanced Software Testing*, vol. 2. Rockynook.

Black, R. (2009). *Managing the Test Process*, 3rd edition. Wiley.

Black, R., van Der Aalts, L., Rommens, J. (2017). *The Expert Test Manager: Guide to the ISTQB Expert Level Certification*. Rockynook.

Boulanger, J.-L. (2012). *Static Analysis of Software*. ISTE-Wiley.

Broekman, B. and Notenboom, E. (2003). *Testing Embedded Software*. Addison-Wesley.

Brooks, F. (2002). *The Mythical Man-Month*. Addison-Wesley.

Buwalda, H., Janssen, D., Pinkster, I. (2002). *Integrated Test Design and Automation*. Addison-Wesley.

Cockburn, A. (2001). *Writing Effective Use Cases*. Addison-Wesley.

Cohn, M. (2006). *Agile Estimating and Planning*. Prentice Hall.

Collard, R. (2002). *Web Testing*. IIST-PSQT/PSTT South Proceedings.

Conklin, A. (2015). *CompTIA Security+ All-in-One Exam Guide*, 4th edition. McGraw Hill.

Craig, R. and Jaskiel, S. (2007). *Systematic Software Testing.* Artech House Publishers.

Crispin, L. and Gregory, J. (2009). *Agile Testing: A Practical Guide for Testers and Agile Teams.* Addison-Wesley.

Crispin, L. and House, T. (2003). *Testing Extreme Programming.* Addison-Wesley.

Demarco, T. and Lister, T. (1999). *Peopleware, Productive Projects and Teams.* Dorset House.

Demarco, T. and Lister, T. (2003). *Waltzing with Bears.* Dorset House.

Deming, D.E. (2000). *Out of the Crisis.* MIT Press.

Dustin, E. (2003). *Effective Software Testing.* Addison-Wesley.

Dustin, E., Rashka, J., Paul, J. (1999). *Automated Software Testing.* Addison-Wesley.

Engel, A. (2010). *Verification, Validation and Testing of Engineered Systems.* Wiley.

Eurocae (n.d.). *DO178 B-ED-12B.* Eurocae.

Eurocae (1999). *Software Considerations in Airborne Systems and Equipment Certification.* Eurocae.

Farrell-Vinay, P. (2008). *Manage Software Testing.* Auerbach Publications.

Fewster, M. and Graham, D. (1999). *Software Test Automation.* Addison-Wesley.

Firesmith, D. (2014). *Common System and Software Testing Pitfalls.* Addison-Wesley.

Ford, N., RIchards, M., Sadalage, P., Dehghani, Z. (2021). *Software Architecture: The Hard Parts.* O'Reilly.

Freedman, D. and Weinberg, G. (1990). *Walkthroughs, Inspections, and Technical Reviews.* Dorset House.

Gärtner, M. (2013). *ATDD by Example.* Addison-Wesley.

Gause, D. and Weinberg, G. (1990). *Are Your Lights On?* Dorset House.

Gilb, T. and Graham, D. (1993). *Software Inspection.* Addison-Wesley.

Goldratt, E. and Cox, J. (2001). *The Goal.* Gower Publishing Ltd.

Goldratt, E.M. and Fox, R.E. (1996). *The Race.* North River Press.

Graham, D. and Fewster, M. (2012). *Experiences of Test Automation.* Addison-Wesley.

Hagar, J.D. (2013). *Software Test Attacks to Break Mobile and Embedded Devices.* CRC Press.

Hendrickson, E. (2013). *Explore IT.* The Pragmatic Programmers.

Homès, B. (2012). *Fundamentals of Software Testing.* ISTE-Wiley.

Homès, B. (2021). *Les tests de logiciels 1: réussir l'ISTQB Fondation.* ISTE Editions Ltd.

Homès, B. (2022). *Advanced Testing of Systems-of-Systems 2: Practical Aspects*. ISTE Ltd and John Wiley & Sons.

Hubbard, D.W. (2008). *The Failure of Risk Management*. Wiley.

Humble, J. and Farley, D. (2011). *Continuous Delivery*. Addison-Wesley.

Humphrey, W.S. (1998). *Managing the Software Process*. Addison-Wesley.

Humphrey, W.S. and Thomas, W. (2010). *Reflections on Management*. Addison-Wesley.

IEEE1028 (2008a). *IEEE 1028 Standard for Software Reviews and Audits*. IEEE.

ISO/IEC/IEEE24748-2 (2018). *ISO/IEC/IEEE 24748-2 Life Cycle Management, Guidelines for ISO/IEC/IEEE15288*.

ISO/IEC/IEEE29148 (2018). *ISO/IEEE 29148 Systems and Software Engineering – Life Cycle Processes – Requirements Engineering*.

ISO/IEC20000-1 (2011). *Information Technology – Service Management – Part 1: Service Management System Requirements*.

ISO/IEC20246 (2017). *IS0/IEC20246 Ingénierie du logiciel et des systèmes – Revue des produits de travail*.

ISO/IEEE (2021). *ISO 24641 Systems and Software Engineering – Methods and Tools for Model-Based Systems and Software Engineering*.

ISO/IEEE29119-3 (2010). *Software and Systems Engineering – Software Testing – Part 3: Test Documentation*.

ISO/IEEE29148 (2018). *Systems and Software Engineering – Life Cycle Processes – Requirements Engineering*.

ISO20246 (2017). *Software and Systems Engineering – Work Product Reviews*. ISO/IEC.

ISO21840 (2019). *ISO 21840:2019 Systems and Software Engineering – Guidelines for Utilization of ISO15288 in the Context of Systems-of-Systems Engineering*. ISO.

ISO25010 (2011). *SQuaRE System and Software Quality Models*. ISO/IEC.

ISO9126-1 (2001). *Quality Model*. ISO/IEC.

ISO9126-2 (2003). *Software Engineering – Product Quality, Part 2: External Metrics*. ISO/IEC.

ISO9126-4 (2004). *Software Engineering – Quality in Use Metrics*. ISO/IEC.

ISTQB (2015). *ISTQB Foundation Level Certified Model-based Tester Syllabus*. ISTQB.

Jones, C. (2000). *Software Assessments, Benchmarks and Best Practices*. Addison-Wesley.

Jones, C. (2007). *Estimating Software Costs*, 2nd edition. McGraw Hill.

Jones, C. (2008). *Applied Software Measurement*, 3rd edition. McGraw Hill.

Jones, C. (2017). *A Guide to Selecting Software Measures and Metrics*. CRC Press.

Jones, C. (2018a). *Quantifying Software, Global and Industry Perspectives*. CRC Press.

Jones, C. (2018b). *Software Methodologies: A Quantitative Guide*. CRC Press.

Jones, C. and Bonsignour, O. (2012). *The Economics of Software Quality*. Addison-Wesley.

Jorgensen, P.C. (2008). *Software Testing: A Craftsman's Approach*, 3rd edition. Auerbach Publications.

Jorgensen, P.C. (2009). *Modeling Software Behavior: A Craftman's Approach*. CRC Press.

Kan, S.H. (2003). *Metrics and Models in Software Quality Engineering*. Addison-Wesley.

Kaner, C. (2013). *The Domain Testing Workbook*. Context Driven Press.

Kaner, C., Falk, J., Nguyen, H. (1999). *Testing Computer Software*, 2nd edition. Wiley.

Kaner, C., Bach, J., Pettichord, B. (2001). *Lessons Learned in Software Testing*. Wiley.

Kim, G. (2019). *The Unicorn Project*. IT Revolution Press.

Kim, G., Behr, K., Spafford, G. (2013). *The Phoenix Project: A Novel about IT, DevOps, and Helping Your Business Win*. IT Revolution.

Kim, G., Humble, J., Debois, P., Willis, J. (2016). *The DevOps Handbook: How to Create World-Class Agility, Reliability, and Security in Technology Organizations*. IT Revolution.

Kneuper, R. (2009). *CMMI Improving Software and Systems Development Processes Using CMMI-DEV*. Rockynook.

Koomen, T.A. (2008). *TMap Next: Pilotage des projets par les tests*. UTN Publishers.

Koomen, T.A., van der Aalst, L., Broekman, B., Vroon, M. (2006). *TMap Next*. UTN Publishers.

Laird, L.E. (2007). *Software Measurement and Estimation: A Practical Approach*. Wiley Interscience.

Linz, T. (2014). *Testing in Scrum*. Rockynook.

Marick, B. (1995). *The Craft of Software Testing: Subsystem Testing*. Prentice Hall.

McConnell, S. (2006). *Software Estimation*. Microsoft Press.

McKay, J. (2007). *Managing the Test People*. Rockynook.

Mosley, D. and Posey, B. (2002). *Just Enough Software Test Automation*. Prentice Hall.

Musa, J.D. (2004). *Software Reliability Engineering: More Reliable Software, Faster and Cheaper*. AuthorHouse.

Myers, G.J. (1979). *The Art of Software Testing*. Wiley.

Nguyen, H., Johnson, R., Hackett, M. (2003). *Testing Applications on the Web: Test Planning for Mobile and Internet-Based Systems*, 2nd proof. Wiley.

Nguyen, H., Hackett, M., Whitlock, B. (2006). *Happy About Global Software Test Automation*. HappyAbout.

Pelisier, C. (1993). *Guide de Sécurité des Systèmes UNIX*. Hermes.

PMI (2017). *PMBOK Guide*, 6th edition. Project Management Institute.

Poppendieck, M. and Poppendieck, T. (2006). *Agile Testing: A Practical Guide for Testers and Agile Teams*. Addison-Wesley.

Poppendieck, M. and Poppendieck, T. (2015). *Lean Software Development: An Agile Toolkit*. Addison-Wesley.

Pugh, K. (2010). *Lean-agile Acceptance Test-driven Development*. Addison-Wesley.

Quigley J.M. and Robertson, K.L. (2020). *Configuration Management: Theory and Application for Engineers, Managers and Practitioners*, 2nd edition. CRC Press.

Riley, T. (2010). *Beautiful Testing*. O'Reilly.

Rothman, J. (2007). *Manage IT!* Pragmatic Bookshelf.

Sacquet, A. and Rochefolle, C. (2016). *Mettre en oeuvre DevOps*, 2nd edition. Dunod.

Savage, S.L. (2012). *The Flaw of Average*. Wiley.

Schwaber, K. (2004). *Agile Project Management with SCRUM*. Microsoft Press.

Schwaber, K. (2014). *The Enterprise and Scrum*. Microsoft Press.

Smith, A.L., Black, R., Davenport, J., Olszewska, J., Rößler, J., Wright, J. (2022). *Artificial Intelligence and Software Testing*. BCS.

Spillner, A., Rossner, T., Winter, M., Linz, T. (2007). *Software Testing Practice: Test Management*. Rockynook.

Splaine, S. (2002). *Testing Web Security: Assessing the Security of Web Sites and Applications*. Wiley.

Splaine, S. and Jaskiel, S. (2001). *The Web Testing Handbook*. STQE Publishing.

Stamatis, D. (2003). *Failure Mode Effect Analysis: From Theory to Execution*, 2nd edition. ASQ.

Syed, M. (2015). *Black Box Thinking: Why Most People Never Learn from Their Mistakes – But Some Do*. Penguin Random House.

Taleb, N.N. (2008). *Black Swann*. Penguin.

Tarantino, A. (2008). *Governance, Risk and Compliance Handbook*. Wiley.

Utting, M. and Legeard, B. (2007). *Practical Model-Based Testing: A Tools Approach*. Morgan Kaufmann.

Walton, M. (1986). *The Deming Management Method*. Penguin.

Watkins, J. (2009). *Agile Testing: How to Succeed in an Extreme Testing Environment.* Cambridge University Press.

Weinberg, G. (2008). *Perfect Software and Other Illusions About Testing.* Dorset House.

Westerman, G.A. (2007). *IT Risk: Turning Business Threats Into Competitive Advantage.* Harvard Business School Press.

Whittaker, J.A. (2003). *How to Break Software.* Addison-Wesley.

Whittaker, J.A. (2009). *Exploratory Software Testing.* Addison-Wesley.

Whittaker, J.A. and Thompson, H. (2003). *How to Break Software Security.* Addison-Wesley.

Wiegers, K.E. (2006). *More About Software Requirements.* Microsoft Press.

Wiegers, K.E. (2007). *Practical Project Initiation: A Handbook with Tools.* Microsoft Press.

Wiegers, K.E. (2010). *Peer Reviews in Software: A Practical Guide.* Addison-Wesley.

Wiegers, K. and Beatty, J. (2013). *Software Requirements*, 3rd edition. Microsoft Press.

Wysopal, C., Nelson, L., Dustin, E., Zovi, D. (2007). *The Art of Software Security Testing.* Addison-Wesley.

Index

L, M, N

O, P, R

S, T, U

V, W, X

Summary of Volume 2

Chapter 5. Test Workload Estimation

Other titles from

iSTE

in

Computer Engineering

2022

MEHTA Shikha, TIWARI Sanju, SIARRY Patrick, JABBAR M.A.
Tools, Languages, Methodologies for Representing Semantics on the Web of Things

SIDHOM Sahbi, KADDOUR Amira
Systems and Uses of Digital Sciences for Knowledge Organization (Digital Tools and Uses Set – Volume 9)

ZAIDOUN Ameur Salem
Computer Science Security: Concepts and Tools

2021

DELHAYE Jean-Loic
Inside the World of Computing: Technologies, Uses, Challenges

DUVAUT Patrick, DALLOZ Xavier, MENGA David, KOEHL François, CHRIQUI Vidal, BRILL Joerg
Internet of Augmented Me, I.AM: Empowering Innovation for a New Sustainable Future

HARDIN Thérèse, JAUME Mathieu, PESSAUX François,
VIGUIÉ DONZEAU-GOUGE Véronique
Concepts and Semantics of Programming Languages 1: A Semantical Approach with OCaml and Python
Concepts and Semantics of Programming Languages 2: Modular and Object-oriented Constructs with OCaml, Python, C++, Ada and Java

MKADMI Abderrazak
Archives in The Digital Age: Preservation and the Right to be Forgotten (Digital Tools and Uses Set – Volume 8)

TOKLU Yusuf Cengiz, BEKDAS Gebrail, NIGDELI Sinan Melih
Metaheuristics for Structural Design and Analysis (Optimization Heuristics Set – Volume 3)

2020

DARCHE Philippe
Microprocessor 1: Prolegomena – Calculation and Storage Functions – Models of Computation and Computer Architecture
Microprocessor 2: Core Concepts – Communication in a Digital System
Microprocessor 3: Core Concepts – Hardware Aspects
Microprocessor 4: Core Concepts – Software Aspects
Microprocessor 5: Software and Hardware Aspects of Development, Debugging and Testing – The Microcomputer

LAFFLY Dominique
TORUS 1 – Toward an Open Resource Using Services: Cloud Computing for Environmental Data
TORUS 2 – Toward an Open Resource Using Services: Cloud Computing for Environmental Data
TORUS 3 – Toward an Open Resource Using Services: Cloud Computing for Environmental Data

LAURENT Anne, LAURENT Dominique, MADERA Cédrine
Data Lakes
(Databases and Big Data Set – Volume 2)

OULHADJ Hamouche, DAACHI Boubaker, MENASRI Riad
Metaheuristics for Robotics
(Optimization Heuristics Set – Volume 2)

SADIQUI Ali
Computer Network Security

VENTRE Daniel
Artificial Intelligence, Cybersecurity and Cyber Defense

2019

BESBES Walid, DHOUIB Diala, WASSAN Niaz, MARREKCHI Emna
Solving Transport Problems: Towards Green Logistics

CLERC Maurice
Iterative Optimizers: Difficulty Measures and Benchmarks

GHLALA Riadh
Analytic SQL in SQL Server 2014/2016

TOUNSI Wiem
Cyber-Vigilance and Digital Trust: Cyber Security in the Era of Cloud Computing and IoT

2018

ANDRO Mathieu
Digital Libraries and Crowdsourcing
(Digital Tools and Uses Set – Volume 5)

ARNALDI Bruno, GUITTON Pascal, MOREAU Guillaume
Virtual Reality and Augmented Reality: Myths and Realities

BERTHIER Thierry, TEBOUL Bruno
From Digital Traces to Algorithmic Projections

CARDON Alain
Beyond Artificial Intelligence: From Human Consciousness to Artificial Consciousness

HOMAYOUNI S. Mahdi, FONTES Dalila B.M.M.
Metaheuristics for Maritime Operations
(Optimization Heuristics Set – Volume 1)

JEANSOULIN Robert
JavaScript and Open Data

PIVERT Olivier
NoSQL Data Models: Trends and Challenges
(Databases and Big Data Set – Volume 1)

SEDKAOUI Soraya
Data Analytics and Big Data

SALEH Imad, AMMI Mehdi, SZONIECKY Samuel
Challenges of the Internet of Things: Technology, Use, Ethics
(Digital Tools and Uses Set – Volume 7)

SZONIECKY Samuel
Ecosystems Knowledge: Modeling and Analysis Method for Information and Communication
(Digital Tools and Uses Set – Volume 6)

2017

BENMAMMAR Badr
Concurrent, Real-Time and Distributed Programming in Java

HÉLIODORE Frédéric, NAKIB Amir, ISMAIL Boussaad, OUCHRAA Salma, SCHMITT Laurent
Metaheuristics for Intelligent Electrical Networks
(Metaheuristics Set – Volume 10)

MA Haiping, SIMON Dan
Evolutionary Computation with Biogeography-based Optimization
(Metaheuristics Set – Volume 8)

PÉTROWSKI Alain, BEN-HAMIDA Sana
Evolutionary Algorithms
(Metaheuristics Set – Volume 9)

PAI G A Vijayalakshmi
Metaheuristics for Portfolio Optimization
(Metaheuristics Set – Volume 11)

2016

BLUM Christian, FESTA Paola
Metaheuristics for String Problems in Bio-informatics
(Metaheuristics Set – Volume 6)

DEROUSSI Laurent
Metaheuristics for Logistics
(Metaheuristics Set – Volume 4)

DHAENENS Clarisse and JOURDAN Laetitia
Metaheuristics for Big Data
(Metaheuristics Set – Volume 5)

LABADIE Nacima, PRINS Christian, PRODHON Caroline
Metaheuristics for Vehicle Routing Problems
(Metaheuristics Set – Volume 3)

LEROY Laure
Eyestrain Reduction in Stereoscopy

LUTTON Evelyne, PERROT Nathalie, TONDA Albert
Evolutionary Algorithms for Food Science and Technology
(Metaheuristics Set – Volume 7)

MAGOULÈS Frédéric, ZHAO Hai-Xiang
Data Mining and Machine Learning in Building Energy Analysis

RIGO Michel
Advanced Graph Theory and Combinatorics

2015

BARBIER Franck, RECOUSSINE Jean-Luc
COBOL Software Modernization: From Principles to Implementation with the BLU AGE® Method

CHEN Ken
Performance Evaluation by Simulation and Analysis with Applications to Computer Networks

CLERC Maurice
Guided Randomness in Optimization
(Metaheuristics Set – Volume 1)

DURAND Nicolas, GIANAZZA David, GOTTELAND Jean-Baptiste, ALLIOT Jean-Marc
Metaheuristics for Air Traffic Management
(Metaheuristics Set – Volume 2)

MAGOULÈS Frédéric, ROUX François-Xavier, HOUZEAUX Guillaume
Parallel Scientific Computing

MUNEESAWANG Paisarn, YAMMEN Suchart
Visual Inspection Technology in the Hard Disk Drive Industry

2014

BOULANGER Jean-Louis
Formal Methods Applied to Industrial Complex Systems

BOULANGER Jean-Louis
Formal Methods Applied to Complex Systems:Implementation of the B Method

GARDI Frédéric, BENOIST Thierry, DARLAY Julien, ESTELLON Bertrand, MEGEL Romain
Mathematical Programming Solver based on Local Search

KRICHEN Saoussen, CHAOUACHI Jouhaina
Graph-related Optimization and Decision Support Systems

LARRIEU Nicolas, VARET Antoine
Rapid Prototyping of Software for Avionics Systems: Model-oriented Approaches for Complex Systems Certification

OUSSALAH Mourad Chabane
Software Architecture 1
Software Architecture 2

PASCHOS Vangelis Th
Combinatorial Optimization – 3-volume series, 2nd Edition
Concepts of Combinatorial Optimization – Volume 1, 2nd Edition
Problems and New Approaches – Volume 2, 2nd Edition
Applications of Combinatorial Optimization – Volume 3, 2nd Edition

QUESNEL Flavien
Scheduling of Large-scale Virtualized Infrastructures: Toward Cooperative Management

RIGO Michel
Formal Languages, Automata and Numeration Systems 1: Introduction to Combinatorics on Words
Formal Languages, Automata and Numeration Systems 2: Applications to Recognizability and Decidability

SAINT-DIZIER Patrick
Musical Rhetoric: Foundations and Annotation Schemes

TOUATI Sid, DE DINECHIN Benoit
Advanced Backend Optimization

2013

ANDRÉ Etienne, SOULAT Romain
The Inverse Method: Parametric Verification of Real-time Embedded Systems

BOULANGER Jean-Louis
Safety Management for Software-based Equipment

DELAHAYE Daniel, PUECHMOREL Stéphane
Modeling and Optimization of Air Traffic

FRANCOPOULO Gil
LMF — Lexical Markup Framework

GHÉDIRA Khaled
Constraint Satisfaction Problems

ROCHANGE Christine, UHRIG Sascha, SAINRAT Pascal
Time-Predictable Architectures

WAHBI Mohamed
Algorithms and Ordering Heuristics for Distributed Constraint Satisfaction Problems

ZELM Martin *et al.*
Enterprise Interoperability

2012

ARBOLEDA Hugo, ROYER Jean-Claude
Model-Driven and Software Product Line Engineering

BLANCHET Gérard, DUPOUY Bertrand
Computer Architecture

BOULANGER Jean-Louis
Industrial Use of Formal Methods: Formal Verification

BOULANGER Jean-Louis
Formal Method: Industrial Use from Model to the Code

CALVARY Gaëlle, DELOT Thierry, SÈDES Florence, TIGLI Jean-Yves
Computer Science and Ambient Intelligence

MAHOUT Vincent
Assembly Language Programming: ARM Cortex-M3 2.0: Organization, Innovation and Territory

MARLET Renaud
Program Specialization

SOTO Maria, SEVAUX Marc, ROSSI André, LAURENT Johann
Memory Allocation Problems in Embedded Systems: Optimization Methods

2011

BICHOT Charles-Edmond, SIARRY Patrick
Graph Partitioning

BOULANGER Jean-Louis
Static Analysis of Software: The Abstract Interpretation

CAFERRA Ricardo
Logic for Computer Science and Artificial Intelligence

HOMÈS Bernard
Fundamentals of Software Testing

KORDON Fabrice, HADDAD Serge, PAUTET Laurent, PETRUCCI Laure
Distributed Systems: Design and Algorithms

KORDON Fabrice, HADDAD Serge, PAUTET Laurent, PETRUCCI Laure
Models and Analysis in Distributed Systems

LORCA Xavier
Tree-based Graph Partitioning Constraint

TRUCHET Charlotte, ASSAYAG Gerard
Constraint Programming in Music

VICAT-BLANC PRIMET Pascale *et al.*
Computing Networks: From Cluster to Cloud Computing

2010

AUDIBERT Pierre
Mathematics for Informatics and Computer Science

BABAU Jean-Philippe *et al.*
Model Driven Engineering for Distributed Real-Time Embedded Systems

BOULANGER Jean-Louis
Safety of Computer Architectures

MONMARCHE Nicolas *et al.*
Artificial Ants

PANETTO Hervé, BOUDJLIDA Nacer
Interoperability for Enterprise Software and Applications 2010

SIGAUD Olivier *et al.*
Markov Decision Processes in Artificial Intelligence

SOLNON Christine
Ant Colony Optimization and Constraint Programming

AUBRUN Christophe, SIMON Daniel, SONG Ye-Qiong *et al.*
Co-design Approaches for Dependable Networked Control Systems

2009

FOURNIER Jean-Claude
Graph Theory and Applications

GUEDON Jeanpierre
The Mojette Transform / Theory and Applications

JARD Claude, ROUX Olivier
Communicating Embedded Systems / Software and Design

LECOUTRE Christophe
Constraint Networks / Targeting Simplicity for Techniques and Algorithms

2008

BANÂTRE Michel, MARRÓN Pedro José, OLLERO Hannibal, WOLITZ Adam
Cooperating Embedded Systems and Wireless Sensor Networks

MERZ Stephan, NAVET Nicolas
Modeling and Verification of Real-time Systems

PASCHOS Vangelis Th
Combinatorial Optimization and Theoretical Computer Science: Interfaces and Perspectives

WALDNER Jean-Baptiste
Nanocomputers and Swarm Intelligence

2007

BENHAMOU Frédéric, JUSSIEN Narendra, O'SULLIVAN Barry
Trends in Constraint Programming

JUSSIEN Narendra
A TO Z OF SUDOKU

2006

BABAU Jean-Philippe *et al.*
From MDD Concepts to Experiments and Illustrations – DRES 2006

HABRIAS Henri, FRAPPIER Marc
Software Specification Methods

MURAT Cecile, PASCHOS Vangelis Th
Probabilistic Combinatorial Optimization on Graphs

PANETTO Hervé, BOUDJLIDA Nacer
Interoperability for Enterprise Software and Applications 2006 / IFAC-IFIP I-ESA'2006

2005

GÉRARD Sébastien *et al.*
Model Driven Engineering for Distributed Real Time Embedded Systems

PANETTO Hervé
Interoperability of Enterprise Software and Applications 2005